Parasite Biodiversity

Parasite Biodiversity

Robert Poulin and Serge Morand

Smithsonian Books
Washington

Copy editor: Fran Aitkens, ELS

Production editor: Joanne Reams

Designer: Brian Barth

Library of Congress Cataloging-in-Publication Data

Poulin, Robert, 1963–

    Parasite biodiversity / Robert Poulin and Serge Morand.

      p. cm.

    Includes bibliographical references and index.

    ISBN 1-58834-170-4 (alk. paper)

    1. Parasites. 2. Species diversity. I. Morand, S. II. Title.

QL757.P69   2004

578.6′5—dc22   2004052536

British Library Cataloging-in-Publication Data available

Manufactured in the United States of America

09 08 07 06 05 04   5 4 3 2 1

∞ The paper used in this publication meets the minimum requirements of the American National Standard for Information Sciences—Permanence of Paper for Printed Library Materials ANSI Z39.48-1992.

Contents

Preface

There is more diversity in nature than the human eye can see. On a coral reef, a diver can spot well over 100 species of fish in less than an hour. Impressive though that number may sound, it is nothing compared with the total number of parasite species that live, unseen, inside those fish or on their external surfaces. This often-overlooked biodiversity is the subject matter of this book.

How many parasite species are there? What causes some host species to harbor many more parasite species than other, related host species? Are there habitats or geographic areas where there are more parasite species per host species than in others? What is so special about the parasite lineages that have diversified at much higher rates than other parasite lineages? And just how important is parasite biodiversity to the evolution and ecology of other organisms? These are the kinds of questions we tackle here to shed more light on all aspects of parasite biodiversity. This book is the first broad synthesis of a growing number of studies currently scattered throughout the scientific literature; by bringing them together, we aim to highlight common patterns and uncover general mechanisms responsible for parasite diversification. Our hope is that the reader, when later gazing at a bird gliding overhead, will see not just the bird, but will also see (or at least imagine!) the flying assemblage of parasite species, representing diverse phyla, that accompanies the bird on its journey through life.

Many of the questions we address apply to a scale larger than that at which most parasitological studies are performed. Some readers might find this disturbing. Others, familiar with the "macro" approaches that have recently become popular in ecology and evolutionary biology, will feel right at home here. Throughout the book, we are not seeking definitive answers to specific questions; instead, we look for tentative answers to some general questions. A sound background in parasitology is therefore not a prerequisite for the reader; the book is targeted at a broad audience, including parasitologists, but also ecologists, evolutionary biologists, epidemiologists, conservation biologists, and wildlife managers.

We received much encouragement and support, in many ways, from several people during the writing of this book. At the beginning, there was Vincent Burke, our enthusiastic editor at Smithsonian Books, who approached us and planted the idea in our heads that we should write this book. Without his initial push, nothing would have happened. Then, the Royal Society of New Zealand provided support in the form of a James Cook Research Fellowship, awarded to Robert Poulin, which allowed uninterrupted periods for writing. At his end, Serge Morand received support from the Institut Français de la Biodiversité. Throughout the project, we both benefited from the intellectual stimulation and enthusiasm of our graduate students and close collaborators. Toward the end, our colleagues Tom Cribb, Gerald Esch, David Marcogliese, Arne Skorping, and Gabriele Sorci made valuable suggestions on various parts of the book; their advice led to many improvements, but we take full responsibility for any remaining errors of fact or interpretation. Last but not least, all along we were warmly supported by our spouses and children, who provided us with the distractions needed to maintain our sanity.

1

The Diversity of Parasites

At a time when our own species is causing profound and global environmental changes, we are becoming acutely aware of the toll these changes are taking on the other species sharing the planet with us. Ironically, we do not have a name for most of the species on Earth, and we do not even know exactly how many living species coexist at the moment: those that go extinct often do so in total anonymity. Most experts agree that we have probably described and named only about 10 to 20% of all living organisms and that the total number of existing species is in the vicinity of 10 million (May 1988; Hammond 1992). The number of species, or species richness, either on the entire planet or in a given area, represents only one component of biodiversity. The term *biodiversity* encompasses all of life's variety, from genetic variation among individuals within a species to variation in community types within a region. Biodiversity is the expression of genetic information at all levels of organization within the biosphere; species diversity is simply its most obvious manifestation and the focus of most biodiversity research. The growing public interest and concern for the preservation of biodiversity have been spurred by some eloquent ecologists (see Wilson 1992 for a stimulating example) who emphasized the practical as well as scientific benefits that would follow from a complete biodiversity survey.

One type of organism often misses out on the biodiversity preservation bandwagon—parasites. Parasites, in fact, are the only organisms targeted by concerted scientific efforts aimed at their partial eradication or total extinction. Drugs, vaccines, and other agents of extinction are aimed at parasites of humans and domestic animals. In relative terms, only a small fraction of existing parasites are of medical or veterinary importance; the rest form an integral part of all ecosystems and we generally ignore them. They do, however, represent a substantial portion of global biodiversity if we adopt the genetic view of biodiversity presented above. Somewhere between 30 and 50% of known animal species can be classified as parasites under a broad definition of parasitism (Price 1980; Windsor 1998; de Meeûs and Renaud 2002). Parasites *sensu lato* include animals that range from plant-eating insects to avian brood parasites. Stricter and more conventional definitions of parasitism, in which a parasite must not only feed on its host without killing it, but also spend a significant portion of its life associated with the host, yield lower estimates of diversity. The total number of parasite species is still likely to be huge, however, because practically all free-living metazoan species harbor at least one parasite species.

Like other zoological disciplines, parasitology suffers from a shortage of well-trained taxonomists and systematists; descriptions of new species and the elucidation of how they relate to previously known species thus progress rather slowly (Brooks and Hoberg 2000, 2001). There are excellent reasons to include parasites in any biodiversity survey, and indeed to study parasite diversity on its own. Parasite diversity provides insights into the history and biogeography of other organisms, into the structure of ecosystems, and into the processes behind the diversification of life (Brooks and Hoberg 2000; Poulin and Morand 2000). Parasites are not to be ignored.

This book is about parasite diversity in all its major aspects. First, we will see if it is possible to estimate parasite diversity in relative and absolute terms. We will examine the patterns in the distribution of parasite diversity among host species and among geographic areas. We will discuss the evolutionary, ecological, and epidemiological processes responsible for the generation and maintenance of parasite diversity. And we will explore the ways in which parasite diversity is important for the rest of the biosphere and the reasons why it should be protected. The book addresses some of the fundamental questions at the core of ecology and evolution, questions that have long puzzled biologists (Hutchinson 1959; May 1988; Groombridge 1992). It also follows in the footsteps of recent work on large-scale, macroecological patterns and processes of biodiversity in nonparasitic organisms (Brown 1995; Rosenzweig 1995; Gaston and Blackburn 2000).

We will mainly emphasize parasite species richness, because it is a convenient, although crude, surrogate for true biodiversity, and because it is the best-studied component of parasite biodiversity. Other, more complex measures of diversity are often used by parasite ecologists (e.g., Cabaret and Schmidt 2001) and ecologists in general (Magurran 1988; Purvis and Hector 2000), but none achieves the universality of species richness. There is also considerable intraspecific variation among parasitic organisms (see Maizels and Kurniawan-Atmadja 2002), but we will not examine parasite diversity at this lower level. We will restrict the taxonomic scope of the book to metazoan parasites of animals, except for a few digressions. These are the classic macroparasites of epidemiological models, that is, parasites in which reproduction occurs via the transmission of free-living infective stages that pass from one host to the next. We adopt this restriction because we are more familiar with metazoan parasites and because they have been the subject of many relevant studies; it must be emphasized, however, that many processes and patterns applying to metazoan parasites may also underpin the biodiversity of other groups of parasites. Metazoan parasites of animals include parasitic helminths (flatworms, roundworms, and others) and arthropods (ticks, lice, fleas, and many crustaceans, among others), as well as several other taxa. These are the standard subjects of classic parasitology textbooks, where information about their general biology and life cycles can be found (e.g., Noble et al. 1989; Roberts and Janovy 1996; Kearn 1998; Bush et al. 2001).

In this opening chapter, we first present some essential background on the evolution and ecology of parasitism. We then address an important question: How good is our knowledge of parasite biodiversity? Much of this book assumes that we possess enough information about parasite biodiversity to look into the processes behind parasite diversification. We therefore begin by validating this assumption.

Origins and Known Diversity of Parasites

Parasites do not represent a monophyletic group: they are an assembly of organisms belonging to separate lineages in which a parasitic mode of life has evolved independently (Zrzavy 2001; de Meeûs and Renaud 2002). Metazoan parasites (*sensu stricto*) include representatives from many phyla (Table 1.1), dispersed across the tree of life (Figure 1.1). How many times has parasitism evolved—that is, how many truly separate lineages of parasites

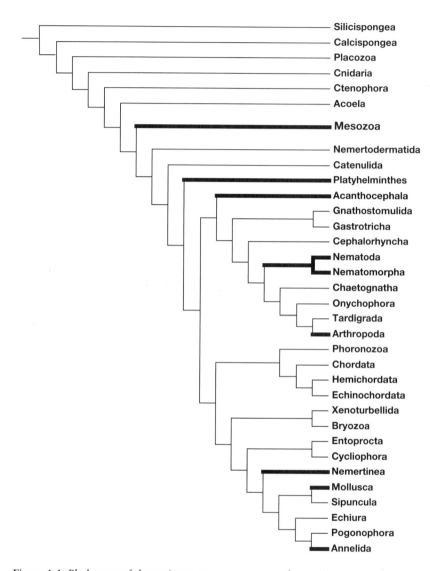

Figure 1.1. Phylogeny of the major metazoan groups and transitions toward parasitism (bold lines). In many phyla (e.g., Nematoda, Arthropoda), transitions to a parasitic mode of life have taken place on several independent occasions. Note: the Pentastomida are grouped within the Arthropoda. (Modified from Zrzavy et al. 1998)

Table 1.1

Minimum numbers of evolutionary transitions to parasitism and numbers of living species in the major groups of metazoan parasites of metazoan hosts (updated from Poulin and Morand 2000)

Parasite Taxon	Minimum Numbers of Transitions	Living Species	Source
Phylum Mesozoa	1	>80	Barnes 1998
Phylum Myxozoa	1	>1,350	Okamura and Canning 2003
Phylum Platyhelminthes*			
Class Cercomeridea (subclasses Trematoda, Monogenea, Cestoidea)	1	>40,000	Brooks and McLennan 1993a; Rohde 1996
Phylum Nemertinea*	1	>10	Barnes 1998
Phylum Acanthocephala	1	>1,200	Amin 1987
Phylum Nematomorpha	1	>350	Schmidt-Rhaesa 1997
Phylum Nematoda*	4	>10,500	Blaxter et al. 1998; Anderson 2000
Phylum Mollusca*			
Class Bivalvia*	1	>600	Davis and Fuller 1981
Class Gastropoda*	8	>5,000	Warén 1984
Phylum Annelida*			
Class Hirudinea*	3	>400	Siddall and Burreson 1998
Class Polychaeta*	1	>20	Hernández-Alcántara and Solis-Weiss 1998
Phylum Pentastomida	1	>100	Barnes 1998
Phylum Arthropoda*			
Subphylum Chelicerata*			
Class Arachnida*			
Subclass Ixodida	1	>800	Klompen et al. 1996
Subclass Acari*	2	>30,000	Houck 1994
Subphylum Crustacea*			
Class Branchiura	1	>150	Barnes 1998
Class Copepoda*	9	>4,000	Humes 1994; Poulin 1995a
Class Cirripedia*			
Subclass Ascothoracida	1	>100	Grygier 1987
Subclass Rhizocephala	1	>260	Høeg 1995
Class Malacostraca*			
Order Isopoda*	4	>600	Brusca and Wilson 1991; Poulin 1995b
Order Amphipoda*	17	>250	Kim and Kim 1993; Poulin and Hamilton 1995
Subphylum Uniramia*			
Class Insecta*			
Order Diptera*	2	>2,300	Price 1980
Order Phthiraptera (suborders Ischnocera, Amblycera, Anoplura)	1	>3,000	Barker 1994
Order Siphonaptera	1	>2,500	Roberts and Janovy 1996

* Taxon also contains free-living species.

are there? This question could be answered precisely only if we had a fully resolved phylogeny of the animal kingdom. For instance, although the entire parasitic phylum Acanthocephala issued from a single ancestral transition to parasitism, more than 10 such transitions have occurred within the order Copepoda alone, a fact only recognized after detailed phylogenetic reconstruction of the evolutionary history of copepods (Poulin 1995a). In addition, the phylogenetic affinities of some large groups of parasites, such as the phylum Myxozoa, remain unclear (Okamura and Canning 2003; Canning and Okamura 2004). Based on current information, we can say that living metazoan parasites of animals are the product of at least 60 independent evolutionary transitions from a free-living existence to one of obligate parasitism.

It is difficult to obtain an accurate estimate of the number of known (i.e., described in the scientific literature) species in each parasite group. Up-to-date compilations of known species are rare. Furthermore, it can be argued that the definitions of what constitutes a species currently used by biologists may not be appropriate for several parasite taxa, and their use may lead to gross underestimates of parasite biodiversity (de Meeûs et al. 2003). It is possible, however, to derive an absolute minimum number of living species based on the approximate number of species described and an expert's opinion on what is known to exist but has yet to be formally described. Table 1.1 presents these minimum numbers. Their sum shows that this book deals with more than 100,000 of the approximately 1.5 million species known to exist. This is a respectable number, but as we will see in the next chapter, one that is probably a gross underestimate of the true parasite biodiversity.

It must be pointed out that most of these species represent single entries in the scientific literature: they are described in one original paper and never mentioned again. We know practically nothing of their biology. Even our descriptions of the species are based on incomplete information. For instance, most copepods parasitic on fish are described based on their females only: males have never been seen. Also, most parasitic helminths have complex life cycles involving two or more hosts. Typically, we know a helminth species by only one of its developmental stages, usually the adult; what the other stages look like or in what host species they are to be found is unknown. Thus our knowledge of parasite biodiversity is often restricted to information on the existence of certain species and their general appearance and whereabouts, and little more.

A Primer in Parasite Ecology

Parasite biodiversity can be studied at several spatial scales. The scales can be defined by the host, in which case they correspond to the host individual, the host population, and the host species; they can also be defined by geographic areas. At each scale, the richness and exact composition of the assemblage of parasite species will be the outcome of different factors. The term *assemblage* is used here to refer to all parasite species found within one specified subdivision of any given scale of study (e.g., all helminth species in the gastrointestinal tract of all individuals of one host population). We generally avoid the related terms *community* and *guild* unless they clearly apply or unless they are incorporated in standard parasite ecology jargon; their use normally implies that the species within them are interacting, and this is not always the case in many parasite assemblages.

The parasite species exploiting a host species can be studied at several hierarchical levels (see Holmes and Price 1986; Bush et al. 1997; Poulin 1998a). These levels are defined by the choice of physical scale for the study of parasite assemblages (Figure 1.2). At the lowest level and smallest scale we have the *infracommunity,* comprising all parasites of different species within the same host individual. Infracommunities are short lived, their maximum lifespan being that of the host. During that time, they are in constant turnover because of the recruitment of new parasites from the pool of locally available species and the death of old ones. Because examination of different host individuals from the same population allows the census of several replicate infracommunities, robust statistical tests of species interactions or deviations from null models of species assembly are possible at this level (e.g., Lotz and Font 1991; Poulin 1996a; Poulin and Guégan 2000; Mouillot et al. 2003). However, because the diversity of parasites within a single host individual is dependent on many factors, from the season of capture to purely stochastic factors, and is not often a reflection of the diversity of the pool of locally available parasite species, we will not focus on this level. For instance, Kennedy and Guégan (1996) found that infracommunities of intestinal helminths in eels, *Anguilla anguilla,* become saturated with parasite species at values well below the number of parasite species locally available. Infracommunities of larval trematodes in gastropod hosts provide an even better illustration: individual snails almost always harbor a single trematode species, whereas several trematode species can coexist in the same snail population (Sousa 1993; Kuris and Lafferty 1994; Esch et al.

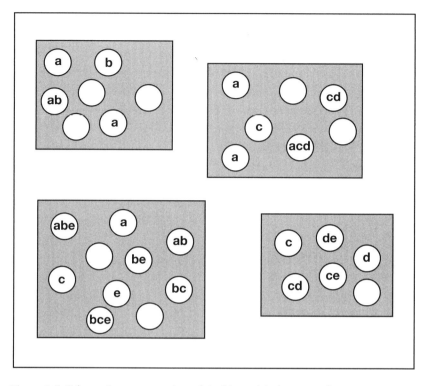

Figure 1.2. Schematic representation of the hierarchical nature of parasite assemblages and the levels at which parasite biodiversity can be studied. The outer box represents the *parasite fauna* of a host species, the grey boxes represent the various host populations, and the white circles represent host individuals. In this case, the parasite fauna includes five parasite species (a to e). Each grey box contains a *parasite component community* that consists of all parasites of all species exploiting that host population; usually, these parasite species are only a subset of those found in the parasite fauna. Within a host population, some host individuals are uninfected, whereas others each harbor an *infracommunity* of parasites.

2001). Clearly, infracommunity richness may be a poor index of local parasite richness.

The next level up is that of the *component community,* which consists of all parasite species exploiting a host population at a given point in time (Figure 1.2). The component community provides the local pool of parasite species from which infracommunities are formed. An analogy with metapopulation biology is evident (see Hanski and Simberloff 1997). Each host individual can be seen as a habitat patch containing subsets (*infrapopulations*) of several metapopulations of different parasite species. Infection by one parasite species is equivalent to patch colonization, and death or recovery of the host is equivalent to local extinction within the patch. The component community is the sum of all parasite metapopulations and thus exhibits the properties of a metacommunity.

The component community is a longer-lived assemblage than the infracommunity; its richness decreases as certain parasite species become locally extinct and increases as others colonize the host population from other, nearby populations. The similarity in species composition between the component parasite communities of different, but conspecific, host populations will depend on their geographic proximity to one another and on the possibilities of parasite exchanges among them (Esch et al. 1988; Poulin and Morand 1999). As with other types of ecological communities (Nekola and White 1999), the similarity between component communities of parasites often tends to decay exponentially with geographic distance (Poulin 2003). Therefore, no single host population (i.e., component community) is likely to include all species of parasites known to exploit the host species; instead, each component community is a subset of a larger collection of species referred to as the *parasite fauna* of the host species. Parasite faunas represent the highest hierarchical level of organization of parasite assemblages for a given host; they are artificial rather than biological entities, but have nevertheless been the subject of many macroecological studies of parasite diversity. In fact, most of the studies discussed later in this book use data from either the component community level or the parasite fauna level; these are the relevant scales for most analyses of parasite diversity patterns among host species.

One caveat in using hierarchical levels based on hosts (individuals, populations, species) to define scales of study in parasite biodiversity is that they consider only one stage in the life cycle of parasites. For instance, gastrointestinal helminths of vertebrates are found as adults in their vertebrate hosts; taken together, the adult helminths of one fish species form an assemblage. During their larval stages, however, these parasites will occur in different host species and will be part of different, perhaps less rich, parasite assemblages. This is not a major problem when the question to be answered relates to the sort of host characteristics that are associated with rich parasite faunas, but it can be a problem if data on parasite species richness are to be used to extrapolate some estimate of the number of parasite species present in an area.

It is also possible to define scales of study based on geography rather than hosts, at local, regional, continental, or similar kinds of spatial scales, and derive lists of parasite species found in different areas regardless of whether they share the same hosts. This approach has sometimes also been used in the context of parasite biodiversity, but not as often as those involving host-based scales. Given that the physical habitat of parasites is nicely delimited by host bodies, at least for most of a parasite's life, host bodies provide a good basis as units of study. Whatever system is used, defining the scale of study is crucial because of its effect on parasite species richness. For instance, the "assemblage of metazoan parasites in North American eels (*Anguilla rostrata*)" is meaningless without an accompanying description of the scale of study. Species richness of metazoan parasites in eels averages about five species per sampling site in rivers of Nova Scotia, Canada; this value is almost doubled if the scale of study becomes the entire watershed instead of the local sampling station, and is more than tripled if the scale of study becomes the entire parasite fauna of eels across North America (Barker et al. 1996). As in everything else in ecology, scale matters in studies of parasite diversity.

How Well Do We Know Parasite Biodiversity?

Obtaining a precise count of living parasite species is currently impossible because we still have not identified all living host species. For instance, in the past 10 years alone, several hundred new amphibian species have been described, and this rate shows no sign of slowing (Glaw and Köhler 1998; Hanken 1999). Clearly, if the list of existing vertebrate species is far from complete, we are a long way from a comprehensive inventory of all free-living species. This is even more true for parasite species for at least two reasons. First, parasites can be found and described only *after* their host species have become known to science. Some larval helminth parasites are often described before their definitive host (in which the adult parasites occur) are identified, but these definitive hosts are previously described species. We can think of only one exception to this rule: for unusual reasons, including mild procrastination, the parasitic copepod *Dinemoleus indeprensus* (Cressey and Boyle 1978) was formally described years before its host, the deep-sea "megamouth" shark (Taylor et al. 1983; Berra 1997). The typical delay, though, creates an inevitable time lag between the eventual completion of host and parasite species lists. Second, parasite species are obviously smaller than their hosts and are therefore more easily overlooked even when the host species is well known. These problems are unavoidable and suggest that the numbers in Table 1.1 are gross underestimates of true diversity.

The bottom line is that we are far from a complete knowledge of parasite diversity. This statement is likely to be more true for some geographic areas, habitats, or host taxa than for others. This disparity in knowledge may result from the work of one or a few influential and highly prolific parasitologists who produce remarkably exhaustive inventories of species in certain taxa or areas compared with the surveys available for other groups or other regions. As a rule, parasite biodiversity is poorly known in tropical areas (e.g., Lim 1998), but there are also gaps in our knowledge of marine parasites in temperate regions (e.g., New Zealand; see Poulin 2004a). The unequal knowledge of different parasite groups can also be a consequence of the biological properties of these taxa, such as size or habitat, which determine how easy they are to observe. For instance, we know more about the diversity of mammals or vertebrates in general than of invertebrates, just as we know more about temperate terrestrial environments than deep-sea habitats. Consequently, our knowledge of mammal or vertebrate parasites from temperate terrestrial zones should be better than our knowledge of the parasites of tropical deep-sea invertebrates.

Is it premature, then, to address questions about the patterns in parasite diversity and the processes underlying those patterns? If our knowledge of parasite biodiversity is still in its infancy, trying to answer such questions would be a waste of time. Do we know enough about parasite diversity to tackle these questions?

To answer this question, we could determine whether the number of new parasite species described per year is gradually decreasing. Given the finite number of species to be described, a slowing down in the rate of new discoveries could indicate that we are approaching the end of our tally. However, the past two decades have seen taxonomy and systematics fall out of favor with funding agencies, and a lower rate of new species descriptions may simply reflect the fact that fewer people are carrying out this sort of work. For instance, the number of new helminth species being described annually in the scientific literature showed a sharp decline in the 1980s (Fig-

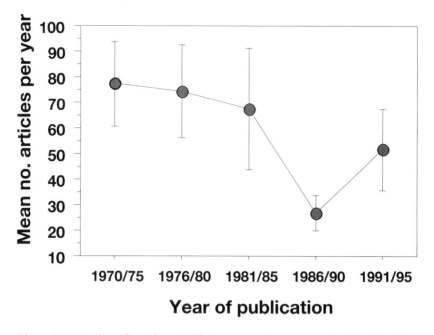

Figure 1.3. Number of articles published per year (mean ± standard deviation) in which the expression "n. sp." appeared in either the title or the abstract, based on a search of the *Helminthological Abstracts*. This provides an estimate of the number of papers describing new helminth species as a function of publication year. (Data from Hugot et al. 2001)

ure 1.3). The subsequent increase during the 1990s is probably due to the application of new molecular and biochemical techniques to the study of helminth systematics. In other words, species complexes are being revealed where it was once thought that there was a single, phenotypically variable species (see Chapter 2). It does not mean that a resurgence of classical taxonomy is taking place, with many entirely new host species or geographic areas being actively surveyed for parasites.

The problems of using the rate of species description as an indicator of diversity become more acute when the focus is on a limited geographic area rather than on a global scale. In Mexico, for instance, the first report of a trematode appeared in the 1930s. Since then, only 25 different people have authored papers on Mexican trematodes (Pérez-Ponce de León 2001). With such a small pool of active taxonomists spread over 70 years, we cannot expect a constant rate of species descriptions. This is also well illustrated by

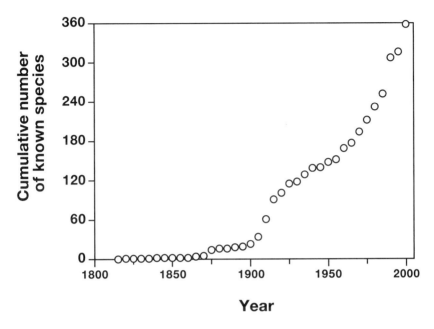

Figure 1.4. Cumulative number of cestode species from Australian vertebrates known to science over time. (Data from Beveridge and Jones 2002)

the rates of discovery of new parasitic cestode species from Australian vertebrates (Figure 1.4). The fauna is clearly still poorly known because there is no hint that the rate of species description is slowing down. More interestingly, a clear hump in the cumulative curve of species descriptions occurred in the early 1900s, suggesting a peak in cestode discoveries at that time (Figure 1.4). In their historical survey, Beveridge and Jones (2002) have shown that this hump is due to the activities of a single taxonomist, T.H. Johnston, who published mainly during that period. Similarly, the number of known trematodes from Australian fishes has increased sharply over the past 25 years, due to the activities of a single research team, still active, which has described approximately one-third of the 411 species reported to date (Cribb 2004). The effects of individual scientists are therefore likely to be very pronounced at local or regional scales. At such scales, if a single enthusiastic individual can influence the shape of the cumulative curve of species descriptions, then this curve may not be reliable for determining whether we have achieved a sufficient knowledge of a parasite fauna.

Another way to assess the extent of our knowledge of diversity in different parasite groups is to examine the relationship between the body size of known species and their date of description. Typically, in animal taxa such as insects, recently described species tend to be smaller than species known for a long time, simply because the probability of detection increases with body size (Gaston 1991a; Gaston et al. 1995). This pattern does not emerge strongly from studies on vertebrates, where factors like the size of the geographic range and relative abundance are better predictors of date of description (e.g., Gaston and Blackburn 1994; Blackburn and Gaston 1995; Reed and Boback 2002). However, in smaller taxa, the dimensions of an organism have a greater effect on the probability of description. Another reason for the late detection of small-bodied species may be the gradual improvement of collection or examination methods, such as microscopy, making small species accessible to scientists. If there is no negative relationship between body size and year of description in a given taxon, we may infer that it is relatively poorly known, since we are not left with only the small-bodied species to describe.

Relationships between year of description and body size have been examined using large compilations of data on parasite species. For example, among monogenean flatworms ectoparasitic on fish, there is a clear negative correlation between parasite body size and the year they were described (Figure 1.5). The scatter of points in the figure suggests that the relationship is due mainly to the sudden inclusion, beginning in the 1950s, of the small-bodied monogeneans (<1 mm long) in the published record. The relationship holds when it is repeated within monogenean families rather than across all species (Poulin 2002), indicating that it is not a taxonomic artifact. Based on current evidence, the smallest size that a monogenean can reach appears to be around 0.15 mm. The lowest size of described species seems to have reached a plateau in recent years (Figure 1.5), and we probably now have a good idea of monogenean biodiversity in general. No doubt many species are yet to be discovered and described—in particular, monogeneans of tropical fishes are still poorly known. Nevertheless, we have achieved a reasonably good knowledge of monogenean diversity, one that at least covers their entire size range.

The same is true of other parasite groups, but not all. Among trematodes parasitic on mammals, a negative correlation exists between the year in which a species was described and its body size; no such relationship exists among trematodes parasitic on either birds or fishes (Poulin 1996b). Simi-

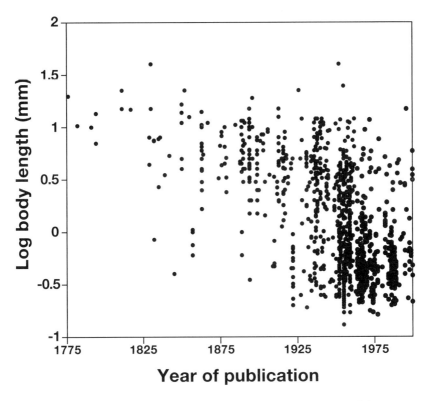

Figure 1.5. Body length of 1,131 species of monogeneans parasitic on fishes as a function of the year in which they were described. (From Poulin 2002)

larly, a negative relationship is found among copepod species parasitic on fishes but not among species parasitic on invertebrates (Poulin 1996b). Clearly, our inventory of some parasite groups is more complete than that of other groups.

Another, related approach to determining whether we have achieved a relatively complete inventory is to examine how either the mean or the mode of the distribution of body sizes in one group of parasites changes over time as more species are described and added to the list of known species. The mode in a frequency distribution is the value represented by the greatest number of individuals; for some types of skewed distributions, it may be a more sensitive measure than the mean. In the case of the body sizes of known species, we would expect both the mean and mode to shift gradually toward smaller body sizes as more species are described. When the

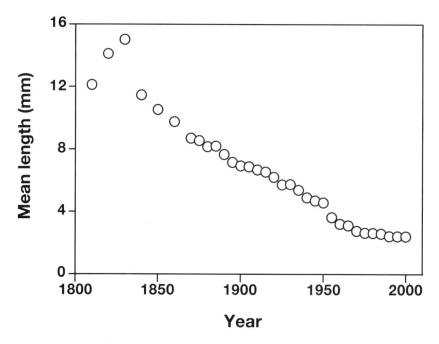

Figure 1.6. Mean body length of known monogenean species over time, showing a decrease in size as more and more species are described. (Data from Poulin 2002)

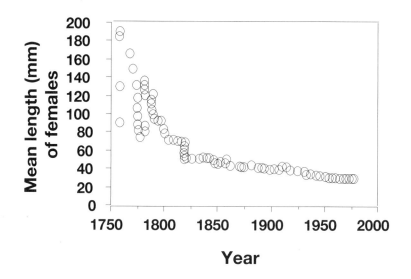

Figure 1.7. Mean body length of females of known species of nematodes parasitic on vertebrates, showing a decrease over time as more species are discovered, of a total of 5,684 species. (Data from Skrjabin 1953–1965)

mean or modal value stops decreasing and levels off, we can be confident that we have covered the full range of body sizes among the existing species in the group, and thus that we have gathered a reasonably representative inventory of existing biodiversity. In monogeneans, the mean body size of known species has decreased gradually, almost linearly, for nearly 150 years, before recently showing signs of stabilizing at around 2.4 mm (Figure 1.6). This suggests that although more species remain to be found, we already have good coverage of the disparity of body sizes among existing species of monogeneans. The same is generally true for other parasite groups, although the details may differ. For instance, in nematode species parasitic on vertebrate hosts, the mean body size of known species initially decreased sharply over time, then much slower, and eventually stabilized at just over 30 mm (Figure 1.7).

Many more parasite species are out there waiting to be discovered and described. Nevertheless, despite the many caveats discussed above, we have a substantial knowledge of parasite biodiversity and it is still possible and informative to investigate patterns in the diversity of parasites as a function of their own biology or that of their hosts.

2

Estimating Parasite Diversity

A test of a scientific hypothesis is only as good as the data on which it relies. Noisy estimates in place of accurate values can lead to unreliable conclusions. In Chapter 1, we saw that parasitologists have made giant strides toward good (although never to be completed) inventories of parasite species, at least in certain geographic areas or for certain host groups. We can therefore make rough guesses at how many species of, say, monogeneans there are on Earth. But to test hypotheses regarding the nature of the processes generating parasite biodiversity or those determining its distribution in space and time, we often need estimates of parasite diversity on finer scales. For instance, we might need to know how many species there are in a particular parasite genus from a given geographic area, or how many parasite species exploit a given population or species of host.

In this chapter, we will examine the problems involved in trying to obtain estimates of parasite diversity at these lower levels. We will also discuss methods proposed to resolve these difficulties. Finally, we will attempt to provide estimates of total species numbers in certain parasite groups at a global level, using approaches commonly applied for estimating arthropod species richness.

Recognizing Parasite Diversity

Imagine a situation where all parasite individuals are sampled from all host species in a habitat or region. In such a situation an accurate tally of parasite species would appear straightforward, but this is not necessarily the case. The problem comes from potential difficulties in distinguishing among different species, difficulties that will lead to erroneous estimates of true species richness.

One possibility is that species richness will be overestimated. A parasite species may exploit more than one host species, but show slight morphological differences related to the host species in which it develops. The phenotype of a parasite is the product of its genotype modulated by its growing environment; genetically identical individuals may thus look slightly different if they develop inside different host species, where living conditions would be different. Considerable host-induced variability in phenotype has indeed been reported in several helminth species (e.g., Kinsella 1971; Blankespoor 1974; Amin 1975; Bray and des Clers 1992; Pérez Ponce de León 1995). This phenomenon has been compounded by the traditional belief that parasites are highly host-specific—often when a parasite genus is found for the first time in a host species, a new species is erected on that basis alone. This is not restricted to adult parasites: Bell et al. (2002) have recently shown that most of the morphological variation among metacercarial stages of two trematode species in the genus *Apatemon* can be attributed not only to the identity of the intermediate host species, but also to the location of the metacercariae within the host, leading them to question the validity of the two distinct species. All this has led to, and continues to lead to (see Johnson et al. 2002b; Valkiunas and Ashford 2002) incorrect splitting of one species into two or more congeneric species, and an inflated estimate of parasite diversity.

It has recently been estimated that more than 25% of named species in diverse but poorly studied taxa will eventually be proven invalid (Solow et al. 1995; Alroy 2002). Formally described species often prove to be synonyms of previously described species (e.g., Dallas et al. 2001; Platt and Jensen 2002) or are invalidated for other reasons. Given the host-induced morphological plasticity of many kinds of parasites, estimates of the species diversity of any higher taxon based on known species may need to be revised downward.

Conversely, species richness can be underestimated: several distinct parasite species can be mistakenly lumped into one. The recent use of molecular approaches in systematics has revealed large numbers of morphologically similar "cryptic" species within taxa previously recognized as a single species (Hillis et al. 1996), and parasite species have been no exceptions (Combes 1995, 2001; Thompson and Lymbery 1996). Several molecular techniques are now available for the taxonomic identification of parasites (MacManus and Bowles 1996; Blouin 2002), and these should prove extremely valuable to studies of parasite biodiversity.

Cryptic species, although highly similar morphologically and often sympatric, are completely isolated reproductive entities and should be regarded as distinct, whatever current definition of a species one cares to choose. For instance, the trematode *Macvicaria crassigula* (Opecoelidae) was known from three congeneric fish in the Mediterranean, *Diplodus sargus, D. vulgaris,* and *D. anularis* (Sparidae). Morphological differences between individual worms from the three host species were too slight to justify splitting them into more than one species (Bartoli et al. 1989). However, a recent comparison of rDNA sequences of worms from all three host species has shown that they represent a complex of two cryptic species, the first restricted to *D. anularis* and the second shared by the other two *Diplodus* species (Jousson et al. 2000). The analysis showed clearly the existence of two well-defined species, whose genetic divergence is typical of that normally observed between congeneric species. The same result was obtained for a second unrelated trematode species, *Monorchis parvus* (Monorchiidae), also previously thought to represent a single species exploiting all three fish species (Jousson et al. 2000). In both cases, there had previously been an error in taxonomy that resulted in two distinct species being lumped under one name. Similar findings have been reported in other trematode taxa (e.g., Reversat et al. 1991; Adlard et al. 1993; Chilton et al. 1999), as well as in other parasite groups such as cestodes (Euzet et al. 1984; Renaud and Gabrion 1988; Ba et al. 1993), monogeneans (Huyse and Volckaert 2002), and nematodes (Chilton et al. 1992; Nascetti et al. 1993; Hoberg et al. 1999; Hung et al. 1999; Feliu et al. 2000; Blouin 2002; Leignel et al. 2002). In most cases, a parasite species once considered capable of parasitizing a few related hosts has proven instead to be a complex of related but distinct species, each highly host-specific.

Molecular analyses do not invariably reveal the existence of sibling species in place of a single species defined by its morphology. For example, three species of trematodes (each originally described based on its appearance) parasitic on coral reef fishes that occur in the west Pacific have been shown to be genotypically uniform across a distance of more than 6,000 km of ocean (Lo et al. 2001). The geographic distance between the most remote fish populations suggested that local genetic divergence could have occurred in their parasites, but rDNA sequences of trematodes from the most distant sites were almost totally homologous. Thus molecular methods can reveal previously unsuspected species complexes, but that does not occur in all situations where they might be expected. Still, the use of molecular tools has generally sharpened our estimates of parasite biodiversity.

Sampling Effects and Extrapolating Diversity

Even when identifying parasite species is not a problem, simply finding them might be. When trying to estimate the species richness of free-living organisms in different communities, ecologists often find that the time spent searching, the number or size of localities visited, or the number of collecting trips all tend to correlate with recorded species richness. The harder you look, the more species you find. Since rare, low-density or well-camouflaged species have a low probability of detection, longer search times or a larger survey area will be necessary to record their presence.

Parasite species are recorded from their presence in hosts, and rarely from the presence of their infective stages in the environment. Thus the effort put in sampling hosts will determine how well we know parasite diversity. This has been recognized for many years. For instance, a quarter-century ago, Kuris and Blaustein (1977) demonstrated clearly that any attempt to use the species richness of ectoparasitic mites on different rodent species to test ecological hypotheses was likely to be flawed because the number of publications on a rodent species correlates strongly with its recorded species richness. If by far the best predictor of parasite species richness on different host species is some measure of how intensely these hosts have been studied, then the role of ecological variables, if any, may be too small to be detected. Whatever measure of host sampling effort is used, there is usually a clear positive relationship between that measure and parasite species richness across different host samples, host populations, or host species (Figure 2.1). The relationship can sometimes be much stronger, such that host sampling effort explains most of the variation in parasite species richness (Gregory 1990; Walther et al. 1995).

Therefore, even in detailed surveys, several parasite species go unrecorded because an insufficient number of hosts are examined. This would not happen if all parasite species exploiting a host population occurred in all host individuals, that is, if all parasite species had a prevalence of 100%. This is not the case: a compilation of published values of the occurrence of para-

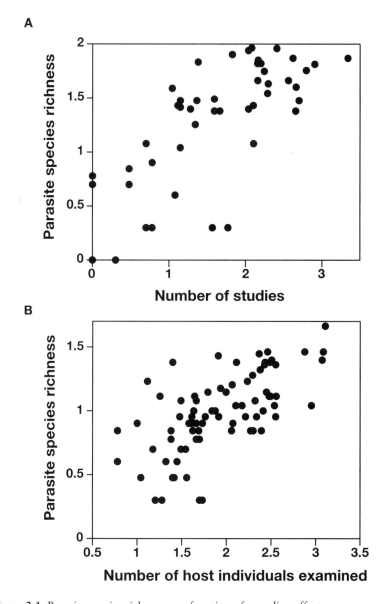

Figure 2.1. Parasite species richness as a function of sampling effort.
(A) Metazoan parasite species richness versus the number of studies published
per host species over a 10-year period, across 49 North American freshwater fish
species (data from Poulin et al. 2000). (B) Gastrointestinal helminth species
richness versus the number of host individuals examined, across 79 mammal
species (data from Morand and Poulin 1998). All data are log-transformed.

sitic helminths in bird and mammal hosts indicates that approximately one-third of parasite species have a prevalence of 5% or less, that is, they occur in less than 5% of host individuals in the population (Figure 2.2). The pattern is not as pronounced among parasites of fishes, but still most species of either external or internal parasites have a prevalence of less than 20% (Figure 2.3). Many of these rare species will remain missing from surveys unless host populations are sampled adequately (Walther et al. 1995). This creates a problem when we wish to compare the species richness of parasite assemblages from different host populations or species, that is, the richness of different parasite component communities. A few host individuals sampled from a species-poor component community, consisting of only highly prevalent parasite species, will yield an accurate estimate of species richness. Conversely, a much larger sample of hosts would be needed to accurately estimate richness in species-rich communities that include several rare parasite species. Without a detailed knowledge of the prevalence values of the different species in the parasite assemblages, how can these estimates be compared?

In comparative studies, two related methods are commonly used to control for the confounding effect of uneven sampling effort when trying to evaluate the independent effect of ecological variables (Walther et al. 1995). First, sampling effort can be included as a predictor variable in a multiple regression model (e.g., Gregory 1990; Gregory et al. 1996). Second, the residuals of a regression of parasite species richness against sampling effort can be used as estimates of richness independent of sampling effort (e.g., Poulin 1995c; Morand and Poulin 1998). Both methods, however, assume that the relationship between the number of species recorded and the number of hosts examined has the same shape in all parasite assemblages (see Walther et al. 1995). This is rarely the case and can lead to biases in the analyses. For each independent host sample, the number of known species in the parasite assemblage increases asymptotically toward the true richness value as more individual hosts are examined (Figure 2.4). The shape and slope of the rising portion of the curve will depend on the prevalence of each parasite species in the assemblage, and will thus differ between different parasite assemblages. This is likely to create some errors when relying on regression methods to correct for uneven sampling effort in comparative studies.

Typically, in such studies, data on parasite species richness, sampling effort, and other variables are culled from the literature; sampling effort

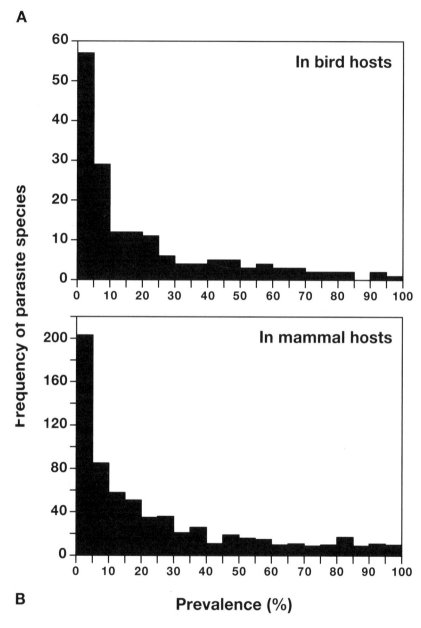

Figure 2.2. Frequency distribution of observed prevalences among 167 species of gastrointestinal helminths from 20 component communities in bird hosts (A), and among 644 species of helminths from 77 component communities in mammal hosts (B). (Data from Poulin 1998b)

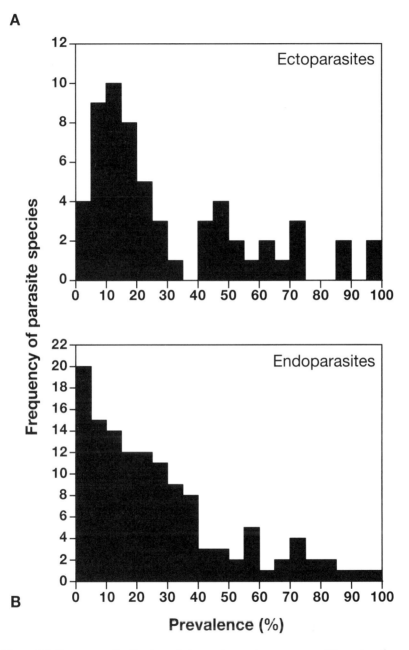

Figure 2.3. Frequency distribution of observed prevalences among 60 species of ectoparasitic metazoans (A), and among 128 species of gastrointestinal helminths (B), obtained from 88 studies of freshwater fish hosts. (Data from Poulin 1998c)

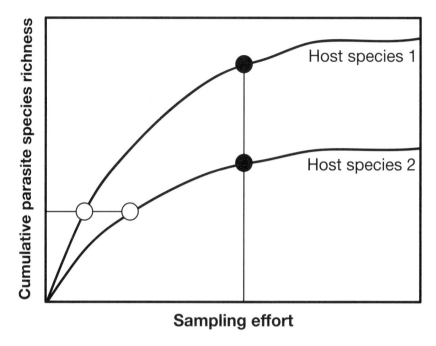

Figure 2.4. Parasite species accumulation curves for two hypothetical host species as a function of sample size. As sampling effort (e.g., the number of host individuals examined) increases, the number of parasite species found approaches the true parasite species richness (the asymptote). For a given sampling effort, more parasite species will be recorded from host species harboring richer parasite faunas (black circles). However, when identical numbers of parasite species are reported from different host species (open circles), the apparent similarity in parasite species richness may be an artifact of unequal sampling effort.

would be recorded as the number of individual hosts examined, and there would be no way of knowing the distribution of parasite species among host individuals. When precise information is available on which parasite species are found in each individual host in a sample, however, a range of extrapolation methods can be used to obtain a richness estimate that includes the rare species missing from the sample. These methods are widely used in studies of communities of free-living organisms (see Palmer 1990; Baltanás 1992; Bunge and Fitzpatrick 1993; Colwell and Coddington 1994). Nonparametric estimators of species richness are easy to compute, and most require only presence–absence data for all observed parasite species in each host examined. From these data, the methods extrapolate the number of probable species missed by inadequate sampling, and add this number to the observed species richness.

Three basic nonparametric methods, as well as modified versions of these same basic methods, have been evaluated specifically for use with estimates of parasite species richness (Poulin 1998b; Walther and Morand 1998). The first method is the (first-order) jackknife estimator, S_j (Burnham and Overton 1979; Heltshe and Forrester 1983):

$$S_j = S_o + a(H - 1)/H$$

where S_o is the observed species richness (the number of parasite species actually occurring in the sample); H is the number of host individuals in the sample; and a is the number of parasite species found in one host in the sample. When $a = 0$, then $S_j = S_o$.

The second method is Chao's (1987) estimator, S_c, which also extrapolates missing species from the number of rare species in the sample:

$$S_c = S_o + (a^2/2b)$$

where b is the number of parasite species found in exactly two host individuals in the sample. Again, when either a or $b = 0$, $S_c = S_o$.

The third method is the bootstrap estimator, S_b (Smith and van Belle 1984):

$$S_b = S_o + \sum_{j=1}^{S_o}[1 - (b_j/H)]^H$$

where b_j is the number of host individuals in the sample in which parasite species j is found. Because even common species contribute to the extrapolation, S_b is always greater than S_o, but only marginally when there are no rare parasite species in the sample.

The performance of these estimators has been evaluated using both real and simulated data (Poulin 1998b; Walther and Morand 1998). Good estimators should be (1) reliable (they should give values close to the true species richness); (2) precise (the values they give should have a low variance); and (3) unbiased (they should not consistently either underestimate or overestimate the true species richness). Tested against real datasets on parasites of vertebrate hosts, the jackknife and Chao estimators proved the least biased and the most precise overall (Walther and Morand 1998). Similar results were obtained using simulated datasets, although the bootstrap method outperformed the others at larger host sample sizes (Walther and Morand 1998) or when many rare species (with low prevalence) are present (Poulin 1998b). For most tests involving either real or simulated data, the three estimators were more reliable than observed parasite species richness as they got closer to the true species richness. From the perspective of biodiversity studies, richness estimators should be used to improve the observed species richness value obtained from a relatively small host sample, that is, to get closer to the true richness value without overshooting it. Since we cannot be certain of the existence of missing species, it may be best to err on the side of caution and settle for an estimate of species richness that is improved while remaining conservative. In this case, the bootstrap method may be preferable (Poulin 1998b), since it reduces the gap between observed and true richness with little chance of overshooting. However, Zelmer and Esch (1999) recommend a higher-order jackknife and argue that the risk of overestimating parasite species richness is no worse than that of underestimating it slightly; this issue is a matter of opinion.

Few comparative studies to date have used parasite species richness estimators instead of observed richness, despite their advantages (e.g., Watve and Sukumar 1995; Morand et al. 2000; Stanko et al. 2002). Richness estimators remain, however, poor substitutes for the actual data in that they offer no clue regarding the taxonomic affinities of missing species, an important piece of information in biodiversity studies. Nevertheless, these estimators suggest that, typically, one or two helminth species escape detection in surveys of vertebrate hosts in which fewer than 40 or 50 individual hosts are examined (Poulin 1998b). Given that sample sizes of vertebrate hosts are often restricted by logistical or ethical constraints to sizes below 50 individuals (Poulin 1998b), we may have incomplete parasite species lists for most host species surveyed to date, in addition to no lists at all for host species yet to be studied.

In light of all this, it is not surprising that the record number of helminth species recorded for a single host species is held by the best known species of all: 287 species of helminths are known to infect humans (Taylor et al. 2001); other estimates extend to well over 300 species (Crompton 1999; see also Ashford and Crewe 2003). No other mammal species comes close to this number, but perhaps only because we have not looked hard enough at their parasites. To do so would come at a major cost, both in animal lives and investigator time. Cribb (1998) has estimated that about 160,000 vertebrates would need to be killed and examined, taking up to 30,000 days of work, just to find all trematode species in Australian vertebrates. Based on the rate at which helminths have been identified from the few Mexican species of amphibians and reptiles examined to date, Pérez Ponce de León et al. (2002) have estimated that it will take more than 660 years to complete the inventory of helminth species in these hosts in Mexico alone.

Extrapolation of Global Parasite Species Richness

Estimating the number of parasite species occurring in a host population is an important first step toward an estimation of global species richness, but it is only a small step. The next step is not an easy one, as illustrated by studies on the world's insect fauna. Since Erwin's (1982) attempt to estimate the number of species of beetles and other arthropods on earth, there have been many criticisms and revisions of these estimates (e.g., Stork 1988; May 1990; Gaston 1991b; Hammond 1992; Gaston and Hudson 1994; Novotny et al. 2002), but we are still in the dark regarding their validity.

Two general approaches are available to estimate global diversity. The first approach is typified by Erwin's (1982) study and was later refined by other researchers (e.g., Ødegaard 2000; Osler and Beattie 2001). Applied to parasite diversity, this method would consist of extrapolating the total number of parasite species on earth, separately for different parasite taxa, using estimates of average species richness per host species, average host specificity, known host diversity, and so on. The second approach consists of extrapolating from rates of taxonomic description, by plotting the cumulative number of species described in a given taxon against time and predicting the asymptote (O'Brien and Wibmer 1979; Dolphin and Quicke 2001; Cabrero-Sañudo and Lobo 2003). There are problems associated with each method. On the one hand, the data needed for the first method are patchy at best, and the different parameters needed for the estimation are perhaps not best characterized by a mean value. An improvement in one of the parameters used to get a richness estimate can completely change the whole estimate (e.g., Novotny et al. 2002). On the other hand, the number of systematists working on a particular parasite taxon will vary through time for many reasons, and there may also be time lags between species discovery and description that would affect the shape of the accumulation curve (see previous chapter). In addition, the potential problem of invalid species mentioned earlier plagues the second method. Still, with these caveats in mind, it is possible at least to attempt an estimation of the number of existing parasite species in different groups.

Using the first approach, we can derive a simple equation to estimate global parasite species richness: (number of host species) times (mean number of parasite species per host species) divided by (host specificity, or mean number of host species exploited per parasite species). Estimates for the latter two parameters are possible, but will be only as accurate as estimates of the first one if the host groups involved are relatively well known, like the vertebrates. Following this approach, Fletcher and Whittington (1998) predicted that there are about 500 species of monogenean parasites on Australian native freshwater fishes, and Whittington (1998) estimated that the world fauna of fishes should harbor approximately 25,000 species of monogeneans, of which no more than 4,000 are described to date. The latter estimate of global monogenean species richness matches that of Rohde (2002), derived independently.

We can apply this approach to other groups of parasites of vertebrates. Three groups of helminths—the trematodes, cestodes, and acanthocephalans—always mature in vertebrates, and thus if we extrapolate the diversity of these groups based on the number of vertebrate species, we can obtain an estimate of global diversity for the parasite groups. Actually, a handful of trematodes mature in invertebrates, but they are so few that they can be ignored here. In the case of nematodes, we can obtain an estimate of the global diversity of the species parasitic on vertebrates, whereas an estimate of their total richness, including the species parasitic on invertebrates, would be more difficult. Using estimates from a range of sources (Table 2.1), we obtained estimates of global trematode, cestode, and acanthocephalan diversity and estimates of the diversity of nematodes parasitic on vertebrates. The numbers we obtained are generally much higher than those presented in Chapter 1 (see Table 1.1) and suggest we have many helminth parasite species yet to find. Estimates of the mean numbers of helminth species per host species used in this analysis are likely to slightly underestimate true richness for two reasons: they are based on component community richness and not parasite fauna richness (see Chapter 1), and they are derived from studies that often focus only on gastrointestinal parasites (some trematodes and nematodes live in other sites). On the other hand, surveys that fail to find helminth parasites from a sample of hosts are probably rarely published, and our estimates may be biased toward higher values than the true averages. Nevertheless, they provide ballpark estimates of global parasite richness that are probably not too far off.

Compared with other estimates of parasite richness based on the same methods but over more restricted geographic areas, the numbers in Table 2.1 may appear rather conservative. For instance, Cribb (1998), using estimates of numbers of parasite species per host species and of host specificity, extrapolated that there are over 6,000 species of trematodes in Australian vertebrates. Using an identical approach, Pérez Ponce de León (2001) estimated that there are more than 8,000 species of trematodes in Mexican vertebrates. Australia and Mexico may be, for whatever reason, hot spots of trematode diversity. Still, the sum of these numbers is more than 50% of the estimate of global trematode species richness presented in Table 2.1. Also, Cribb et al. (2002b) have extrapolated that there are 25,000 to 50,000 trematode species in the world's fishes alone. In light of these other estimates, our values are clearly not exaggerated.

We could normally use the second method mentioned above, the extrapolation from rates of taxonomic description, to validate the numbers in Table 2.1. However, in the case of these parasites, this may be a bit premature. Cumulative numbers of known species plotted against time usually follow a sigmoidal function. In the earliest phase of species discovery, things proceed slowly because no taxonomic framework is in place to accommodate the newly found species; this is followed by a rapid increase in rates of species description, and then eventually by a slowing down when few species are left undiscovered. Extrapolations from current rates of description that do not take into account the slowing down and the asymptote can provide exaggerated estimates of total richness (e.g., Pérez Ponce de León et al. 2002). The choice of mathematical model to fit the data and obtain the asymptote is a matter of debate (Soberón and Llorente 1993; Dolphin and Quicke 2001), but usually all models give roughly similar results, provided that the sigmoidal-like curve is advanced enough. In the case of nematode species parasitic on vertebrates, the inflexion point of the curve has barely been reached, the slowing down is happening right now (Figure 2.5). This makes extrapolation more difficult. The asymptote can be estimated at somewhere between 8,000 and 12,000 species. This is well below the almost 24,000 species extrapolated using the estimates of species richness per host species and host specificity (Table 2.1). For acanthocephalans, the cu-

Table 2.1

Extrapolation of global species richness in parasitic trematodes, cestodes, nematodes, and acanthocephalans

	Chondrichthyes	Osteichthyes	Amphibia	Reptilia	Aves	Mammalia	Total
Known number of host species[a]	843	18,150	4,975	6,300	9,040	4,637	43,945
Mean no. parasite species/host species[b]							
Trematoda	0.12	2.04	1.27	1.06	3.24	1.61	
Cestoda	2.71	1.57	0.27	0.39	3.67	1.89	
Acanthocephala	—	1.01	0.19	0.42	0.72	0.28	
Nematoda	0.48	1.49	2.82	2.15	3.32	3.90	
Mean host specificity[c]							
Trematoda	2.00	6.35	5.40	1.77	2.97	2.01	
Cestoda	1.69	6.38	4.75	2.21	2.36	1.89	
Acanthocephala	—	14.95	6.74	12.50	8.35	4.32	
Nematoda	2.67	10.28	5.27	2.12	3.28	6.07	
Estimated global species richness							
Trematoda	51	5,831	1,170	3,773	9,862	3,714	24,401
Cestoda	1,352	4,466	283	1,112	14,058	4,637	25,908
Acanthocephala	—	1,226	140	212	779	301	2,658
Nematoda	152	2,631	2,662	6,389	9,150	2,979	23,963

[a] From Hammond (1992) and Hugot et al. (2001).

[b] From data of Bush et al. (1990), Poulin (1995c), Curran and Caira (1995), and Goldberg and Bursey (2000), and/or references therein.

[c] From data in Margolis and Arthur (1979), Aho (1990), Bush et al. (1990), Gregory et al. (1991), Poulin (1998c), and Goldberg and Bursey (2000), or references therein.

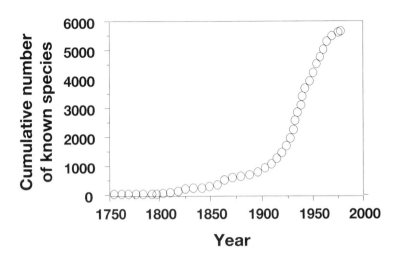

Figure 2.5. Cumulative number of valid species of nematodes parasitic on vertebrates known to science over time. (Data from Skrjabin 1953–1965)

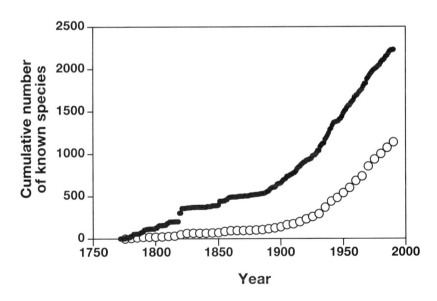

Figure 2.6. Cumulative number of acanthocephalan species known to science over time. The top line includes all proposed species and includes synonyms; the bottom line includes only valid species. (Data from Golvan 1994)

terminants of diverse parasite assemblages in evolutionary time are also important, and will be examined in Chapter 4. Together, this and the next chapter explore the potential causes of variation in parasite species diversity among related host species.

The Parasites' Basic Reproductive Number, R_0

In theory, when a parasite is introduced into a host population, it can follow one of two trajectories: it can go extinct locally or it can spread through the host population until host and parasite reach equilibrium population densities. Epidemiological models allow us to derive a measure of parasite invasiveness, the basic reproductive number, R_0, on which the fate of the parasite depends (Anderson and May 1991; Diekmann and Heesterbeek 2000). For microparasites (bacteria, viruses, protozoans, and fungi) that are capable of rapid multiplication in a host, R_0 is formally defined as the number of secondary infections produced by the first individual pathogen to enter a population of fully susceptible hosts. For macroparasites (helminths and arthropods) where reproduction occurs via the transmission of free-living infective stages that pass from one host to the next, R_0 is defined as the average number of female offspring produced throughout the lifetime of a female parasite that would themselves achieve reproductive maturity in the absence of density-dependent constraints. When R_0 is less than 1, the parasite cannot maintain itself and declines to local extinction. When R_0 is greater than 1, however, the parasite can successfully invade the host population and its numbers grow until equilibrium is reached. In certain contexts, R_0 can also be viewed as a measure of parasite fitness (e.g., Frank 1996).

The exact mathematical expression for R_0 is obtained from the set of differential equations that describe the host and parasite population dynamics, and thus depends on the type of parasite under consideration (e.g., directly transmitted parasites versus parasites with complex life cycles). Mathematical expressions of R_0 therefore vary, but are all formulated as a ratio. In this ratio, host population density and some measure of parasite transmission efficiency (e.g., a birth rate of transmission stages for macroparasites) are in the numerator, and natural host mortality plus parasite-induced host mortality are in the denominator. For instance, let us consider a hypothetical assemblage of directly transmitted helminth parasite species in which adult worms live in the host gut, pass out their eggs in

Table 3.1

Definition of symbols used in calculating the basic reproduction number, R_0 (from Roberts et al. 2002)

Symbol	Definition
α	Increase in host mortality caused by the presence of a single parasite (yr^{-1})
β	Rate at which parasite larvae are eaten by hosts (ha^{-1} yr^{-1})
λ	Rate of egg production by a single adult parasite (yr^{-1})
μ	Rate of adult parasite mortality (yr^{-1})
ρ	Rate at which parasite larvae are lost from the system due to other environmental causes (yr^{-1})
d	Rate of host mortality in the absence of parasites (yr^{-1})
p	Probability that an ingested larva becomes an adult parasite
q	Probability that a parasite egg hatches into a larva
H	Coefficient governing the saturation of parasite transmission (= ρ/β)
K	Host population carrying capacity (ha^{-1})
N	Host population density (ha^{-1})

host feces, and larvae hatched from those eggs infect new hosts that ingest them accidentally. Such helminth assemblages are common in wild grazing mammals. For a single parasite species, we derive (Roberts et al. 2002):

$$R_0 = \left[\frac{K}{H + K}\right] \left[\frac{pq\lambda}{\mu + \alpha + d}\right]$$

The symbols are defined in Table 3.1. This expression for R_0 is valid only for the first parasite species to invade a previously unexploited host population. Most host populations usually harbor assemblages of different parasite species. If n parasite species coexist within a host population in a stable equilibrium, a new species can only successfully invade (and push the parasite species richness to $n + 1$) if $R_0 > 1$ for that species. However, R_0 must now be calculated at the host population density when it is in equilibrium with the pre-existing parasite assemblage (Roberts et al. 2002):

$$R_0 = \left[\frac{N}{H + N}\right] \left[\frac{pq\lambda}{\mu + \alpha + d}\right]$$

Note that K has now been replaced by N, but that nothing else changes in the equation. This apparently minor change is important as it sets the correct conditions under which R_0 must be computed for a new species invading an established parasite assemblage. Since the conditions that determine whether a new species can invade an existing assemblage of parasite species are imposed by the assemblage itself, the order of invasion of different species is also important. Similar expressions of R_0 could be derived for parasite assemblages involving species with life cycles other than direct, or even for an assemblage of parasite species showing a mix of life cycles.

The invasibility of an ith parasite species in a host population can be illustrated using the model of Roberts et al. (2002). An ith parasite species can spread into a host population only if its $R_{0i} > 1$. The value of R_{0i} depends on the host equilibrium density, which depends on the regulation effect of the other parasite species already established in the host population. It depends also on the virulence of the new parasite, on current environmental conditions, and on the parasite transmission rate. For a given set of parameter values, it can be shown that low rates of parasite transmission (β, see Table 3.1) or high virulence may limit the richness of the parasite community by preventing the establishment of new species (Figure 3.1). This becomes exacerbated when several species of parasites already coexist in a host population because of their combined regulatory effect on the host population (Figure 3.2). A more robust investigation of the model is necessary, but this simple example suggests that highly virulent parasites with low transmission rates will have more difficulty spreading into a host population that is already infected by a rich community of parasites.

In the above mathematical expressions, some of the parameters that influence R_0 correspond to intrinsic properties of the parasite species, such as its fecundity (λ) or natural mortality rate within the host (μ). Other parameters, like H and its components ρ and β, depend on the interaction between the host, the parasite, and environmental conditions. Most importantly, however, the expression for R_0 includes intrinsic host characteristics that will affect parasite transmission, such as host population density (N) or natural mortality rate (d). These host features determine to a great extent whether a given parasite species can invade the host population, and whether successive parasite species will also succeed and become established.

The salient point of the above discussion is this: all else being equal, we would expect that interspecific variation among hosts in these features should affect the likelihood that various parasite species can achieve thresh-

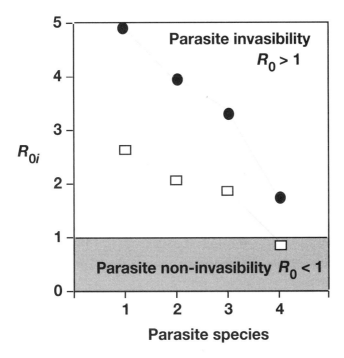

Figure 3.1. Invasibility ($R_0 > 1$) of an ith parasite species, where $i = 1$ to 4, based on the framework of Roberts et al. (2002). Parameter values: host population growth rate = 1.5; host mortality rate $d = 0.01$; parasite egg production rate $\lambda = 6$; adult parasite mortality rate $\mu = 0.1$; larval parasite mortality rate $\mu_1 = 0.1$; host population carrying capacity $K = 100$; coefficient of parasite aggregation among hosts $k = 1$. The virulence α of Parasite Species 1 is 0.001, of Parasite 2 is 0.005, of Parasite 3 is 0.01, and of Parasite 4 is 0.1. The parasite transmission rate β is either 0.0001 (black circles) or 0.00005 (open squares).

old R_0 values (i.e., $R_0 \geq 1$) in these hosts. Host species with features favoring high R_0 values should be easier to colonize by new parasite species, and should therefore harbor richer parasite faunas. For instance, host species occurring at high population densities should be easier to invade for many parasite species, by promoting higher R_0 values, than host species with typically low population densities (Figure 3.3). This and other predicted patterns in parasite species richness issued from epidemiological models tend to be qualitative rather than quantitative, that is, we can predict the direction of a relationship but not its strength or magnitude. The reason for this is that

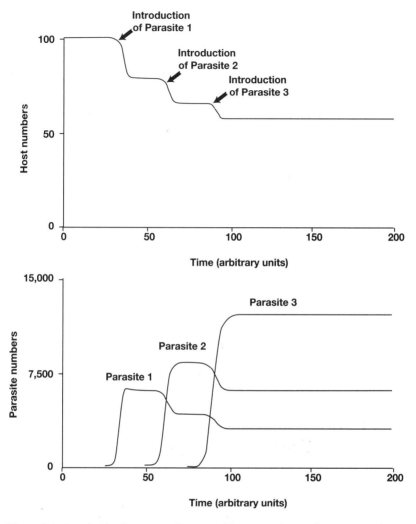

Figure 3.2. Population dynamics of hosts and parasites during the invasion of a host population by three parasite species. In this hypothetical example, each new parasite has a transmission rate β higher than that of previously established species. Host population size (or density) at equilibrium declines with each new parasite species, so that successful invasion becomes more difficult for new parasites unless their basic characteristics (transmission rate, fecundity, etc.) facilitate their establishment.

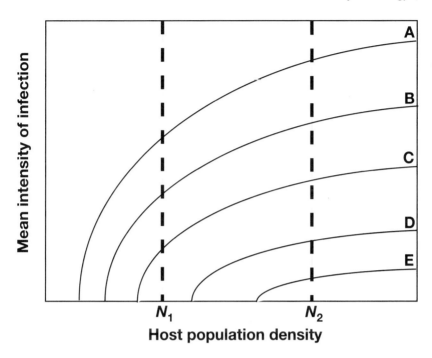

Figure 3.3. The influence of host population density on the intensity of infection (mean number of parasite individuals per host) by five species of parasites, A to E, in the same host population. Parasite species are ranked in decreasing order of R_0 for a given host population density, with Species A having the highest value of R_0 and Species E the lowest. The curve for each parasite species intercepts the x axis at the threshold host density required for that species to be maintained in the host population. Parasite species richness should thus increase with host population density (e.g., from three species at host density N_1 to five species at density N_2). (Adapted from Dobson 1990)

full information would be required for all epidemiological parameters for both hosts and parasites in an association, and for all host and parasite species, to extrapolate a full quantitative relationship. Nevertheless, epidemiological theory provides several testable predictions regarding patterns in parasite diversity.

Empirical Support for Epidemiological Modelling

Before looking at the empirical evidence for relationships between parasite species richness and some key epidemiological parameters, it is worth discussing briefly the relevance of epidemiological models to parasite ecology. How successful have these models been at predicting basic ecological phenomena such as the observed prevalence of parasites among their hosts? Morand and Guégan (2000) addressed this question using data on nematodes parasitic on mammals. They showed that the relationship between mean abundance (mean number of parasite individuals per host) and its variance followed Taylor's power law (Taylor 1961; Taylor et al. 1978):

$$\log(V) = b\log(M) + \log(a)$$

where M is mean abundance and V its variance, a is a constant, and b is an index of aggregation. In comparing large numbers of parasite species, this relationship applies not only to nematode parasites of mammals (Morand and Guégan 2000), but to all types of parasites (see Shaw and Dobson 1995). The linear relationship is pronounced, as seen with data on ectoparasites of vertebrates (Figure 3.4). From basic epidemiological models (Anderson and May 1985), we can link the prevalence of infection to the mean abundance of parasites at any time during an infection according to the following relationship:

$$P = 1 - (1 + M / k)^{-k}$$

where P is the prevalence and k is the aggregation parameter of the negative binomial distribution (Anderson and May 1985). The parameter k relates to the parameters a and b of Taylor's power law (above) as follows:

$$1/k = aM^{(b-2)} - (1/M)$$

These theoretical relationships derived from epidemiological models allow us to compare the prevalences of parasite species observed in nature with those predicted by the models. Across several species of monogeneans ectoparasitic on fishes and fleas ectoparasitic on mammals, the congruence between observed and predicted values is generally quite good (Figure 3.5). It is clear that the models do not capture all of the subtleties of host–parasite dynamics, as seen for instance by the predicted prevalences of fleas on mammals being consistently lower than observed values (Figure 3.5). Still, these results indicate that some fundamental properties of natural host–parasite associations can be predicted by epidemiological models; we can now see whether this applies to parasite species diversity as well.

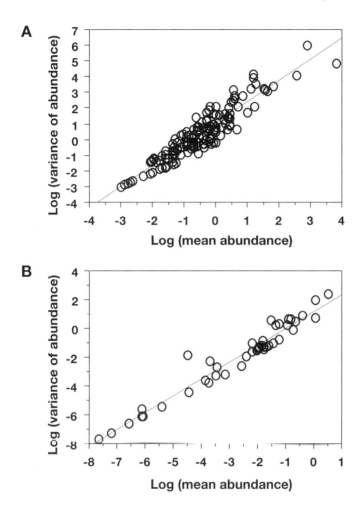

Figure 3.4. Relationships between the variance and the mean abundance of parasites per host among parasite species: (A) *Dactylogyrus* spp. monogeneans on cyprinid fish (data from Simkova et al. 2002); (B) fleas on mammals (data from Stanko et al. 2002).

R_0, Host Features, and Parasite Species Richness

From the above, it is predicted that an increase in host population density will result in an increase in the basic reproductive number, R_0, for all parasite species in contact with the host population. The probability that the infective stages of parasites will contact hosts is assumed to increase with

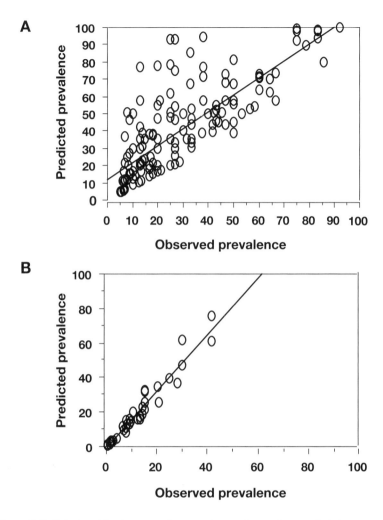

Figure 3.5. Relationships between the prevalence of parasites on their hosts predicted by epidemiological models and the observed prevalence among parasite species: (A) *Dactylogyrus* spp. monogeneans on cyprinid fish (data from Simkova et al. 2002); (B) fleas on mammals (data from Stanko et al. 2002).

the number of hosts within a given area, providing a link between host population density and parasite transmission rate (Anderson and May 1978). There is empirical support for this assumption. Arneberg et al. (1998) found a positive relationship between the abundance of strongylid nematode parasites and host population density across species of mammal

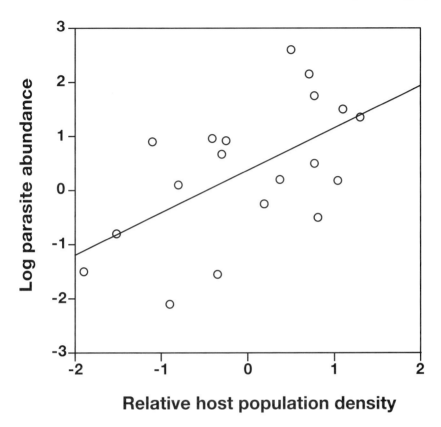

Figure 3.6. Relationship between host population density and abundance (numbers per host individual) of strongylid nematodes across 19 species of mammals. To correct for the effect of body mass, host density is plotted as residuals from a regression of density against host body mass, using log-transformed data; parasite abundance is the within-host average of all strongylid nematodes. (Data from Arneberg et al. 1998)

hosts (Figure 3.6). Host species living at higher population density for their body size generally harbor higher numbers of nematodes (per individual host) than comparable host species living at lower densities. Similarly, Arneberg (2001) reported a positive correlation between the prevalence of oxyurid nematode parasites and host population density across species of mammalian hosts, after controlling for confounding variables. Thus, as host population density increases, a larger proportion of individual hosts are infected by oxyurids. It is worth noting that both strongylid and oxyurid nematodes have direct life cycles, not requiring intermediate hosts. Relationships between host density and measures of parasite transmission success can become obscured in parasite taxa with complex life cycles (Arneberg 2001) because the densities of several hosts would need to be taken into account. Still, there is enough evidence to accept the general link between host population density and parasite transmission rate.

What about parasite species richness? Is it positively related to host population density as it is to parasite abundance and prevalence? If high host density facilitates parasite transmission and allows high R_0 values, it should do so for all locally available parasite species and consequently promote their coexistence in the host population. Few attempts have been made to test this prediction, but they generally support it. After controlling for confounding variables such as host body mass, sampling effort, and phylogenetic relationships among host species, a significant positive correlation was found between host population density and the number of gastrointestinal helminth species harbored by mammalian hosts (Morand and Poulin 1998). This pattern becomes much stronger when the analysis is limited to directly transmitted strongylid nematodes (Arneberg 2002). Similarly, the species richness of fleas on mammals also increases with host population density, across several host species (Stanko et al. 2002). Perhaps the most compelling evidence to date is that provided by a recent comparative study of parasite species richness in primates (Nunn et al. 2003a). Across more than 100 species of primates, Nunn et al. (2003a) found that, of the many host features investigated, host population density was the best predictor not only of total parasite species richness but also of the species diversity of helminths and protozoans taken separately. The link between host population density and parasite diversity is not limited to mammalian hosts. Supporting evidence also comes from a comparative study of endoparasite species richness in chaetodontid fish species on coral reefs (Morand et al. 2000), and from a study of the species richness of monogenean ectoparasites among freshwater fishes (Simkova et al. 2002). The relationships between host population density and parasite species richness in these studies on contrasting host–parasite systems are not always strong (see Figures 3.7 and 3.8), but the fact that they emerge from analyses that control for a range of confounding factors suggests that host density does promote the accumulation of parasite species in a host population. Additional support comes from the observation that tree density correlates strongly with the richness

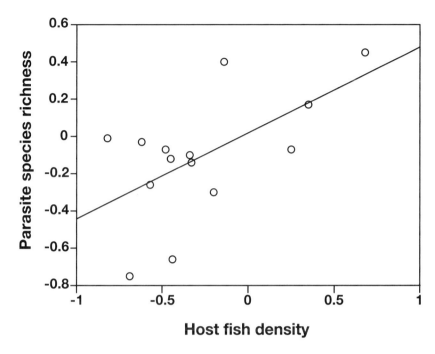

Figure 3.7. Relationship between the species richness of endoparasitic helminths and host density across species of chaetodontid fishes. Points are phylogenetically independent contrasts based on log-transformed data and have been corrected for confounding variables. (Data from Morand et al. 2000)

of phytophagous insects across British and German tree species (Kelly and Southwood 1999; Brandle and Brandl 2001); phytophagous insects are parasitic on trees, and their transmission also depends on the local availability of tree hosts. Of all the published studies to date that have tested this idea, only one study, on chewing lice of Peruvian bird species, reports no association between host population density and parasite species richness (Clayton and Walther 2001).

In the preceding paragraphs, host population density was treated as an average value for the entire population. In most host species, however, the spatial distribution of individuals within the population will be heterogeneous, with high-density patches of hosts separated by areas of low host density. This is typical of many social or gregarious animals. Permanent or long-lived social groups can serve to increase host density on a scale smaller

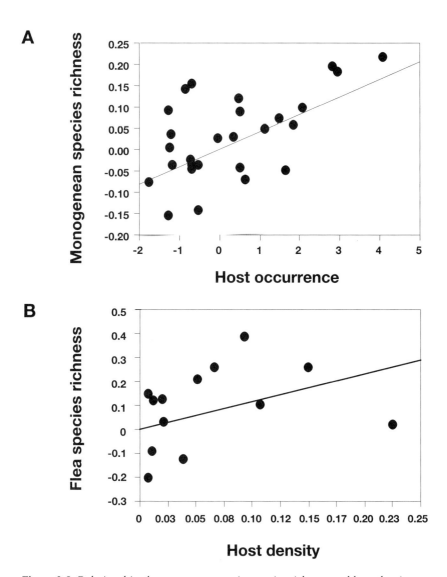

Figure 3.8. Relationships between ectoparasite species richness and host density. (A) Monogenean species richness versus the frequency of occurrence of the host in different localities (a measure of host population abundance), across freshwater fish species (data from Simkova et al. 2002). (B) Flea species richness versus the population density of the host species, across species of mammals (data from Stanko et al. 2002). Points are phylogenetically independent contrasts based on log-transformed data.

than that of the entire population and promote parasite transmission (Altizer et al. 2003). Based on the requirements for the establishment and maintenance of parasite species ($R_0 > 1$) and the influence of host density on R_0, we might expect that all else being equal, gregarious host species will harbor more parasite species than related but solitary host species. There is considerable evidence showing that the prevalence or intensity (e.g., Brown and Bomberger Brown 1986; Moore et al. 1988; Côté and Poulin 1995) and even the diversity (Freeland 1979) of directly transmitted parasites increase with host group size within host species, that is, among distinct groups of conspecifics that vary in membership size but live in the same habitat. Among related host species, however, comparative studies have been limited to fish hosts (e.g., Poulin 1991; Ranta 1992; Caro et al. 1997; Raibaut et al. 1998; Sasal and Morand 1998; but see Poiani 1992), and their conclusions are generally but not consistently supportive. There are good epidemiological reasons to expect an association between host sociality and parasite species richness, at least for certain types of parasites (Altizer et al. 2003), and this potential relationship requires further study.

Depending on the transmission ecology of the parasites, the total number of host individuals available in the host population may be more important for R_0 than the number of hosts per unit area (i.e., host density). In some circumstances, the probability of contacting a suitable and susceptible host may covary more strongly with host numbers, such that in the equation for R_0, the parameter N should be viewed as host population size rather than host population density. In the field, estimates of host population size are rarely available and usually hard to obtain, but surrogate measures are sometimes possible. In a field study of numerous rock-pool populations of the crustacean *Daphnia magna*, water volume of the rock pool was used as an estimator of host population size and was found to correlate positively with endoparasite species richness (Ebert et al. 2001). Similarly, large lakes are inhabited by larger fish populations; among these populations, lake size can correlate positively with parasite species richness (Kennedy 1978; Hartvigsen and Halvorsen 1993). Clearly, confounding factors can explain these results, such as the higher probability that parasites colonize a larger water body than a small one (see Chapter 4), and for this reason the role of host population size *per se* remains less clear than that of host population density.

Host density adopts a different meaning for parasites capable of infecting several coexisting host species. The total density of all suitable hosts, whether conspecifics or not, is what should matter to such parasites, assuming that all hosts are equally suitable. This last assumption may not be met in most natural systems, where different host species, even if closely related, will differ in their ability to mount an immune response against parasites or vary in their suitability in other ways. We are aware of only one study comparing parasite species richness across localities where the assortment of host species ranged from dominance by one or a few species to more equal mixture of several species. Schmid-Hempel (2001) reports that as the diversity of bumblebee host species (*Bombus* spp.) increases, the richness of parasite species exploiting them decreases (Figure 3.9). One explanation consistent with this observation is that it may be more difficult for parasites to invade and persist in diverse host communities, possibly because the highly suitable hosts become diluted among other, less suitable host species. This study did not look at bumblebee density *per se*, only at the diversity of bumblebee species per sampling locality; still, it suggests that hosts of different species are not equal in the eyes of parasites, and that it is only the density (or gregariousness) of the main or most suitable species that links directly with epidemiological predictions.

Host population density is only one of two host features in the equation for R_0 that can influence the magnitude of R_0; the other feature is d, the natural mortality rate of the host population due to causes other than parasitism. Host species in which individuals experience low rates of mortality are the ones in which the probability of surviving from one year to the next is high, and are thus long-lived species. All else being equal, parasite species that invade a population of long-lived hosts are more likely to succeed and become established because their R_0 value will be higher than if the host were short-lived. The prediction issued from the equation for R_0 is therefore that lifespan and parasite species richness should correlate positively across related host species, if there are no major differences in other parameters. The tests of this prediction are not favorable to date. Among freshwater fish species, there is a positive relationship between endoparasitic helminth species richness and host lifespan (Figure 3.10), but this relationship disappears once other variables are accounted for in a multivariate analysis (Morand 2000). Among mammal species, there is actually a

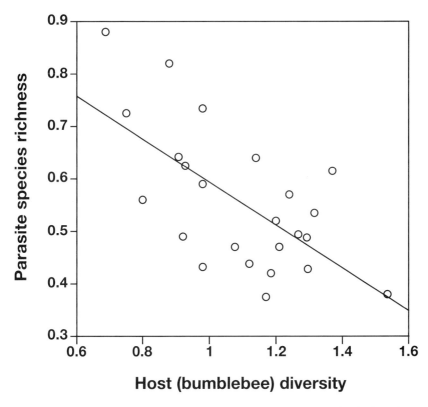

Figure 3.9. Relationship between parasite species richness (per individual worker) and the diversity of the bumblebee (*Bombus* spp.) host community, among different localities in Switzerland. Host diversity is estimated using the Shannon-Wiener diversity index, and parasite species include parasitoid insects, nematodes, mites, and protozoans. (Data from Schmid-Hempel 2001)

negative correlation between host lifespan and helminth species richness, which may be a by-product of other covariates (Morand and Harvey 2000). Clearly, the influence of host lifespan *per se* is difficult to disentangle from that of related life-history variables, and for the moment appears not to be a factor. Host population density, on the other hand, despite also covarying with a constellation of other host features such as body size or metabolic rate, emerges from multivariate analyses as an important determinant of parasite species richness, as expected from epidemiological theory.

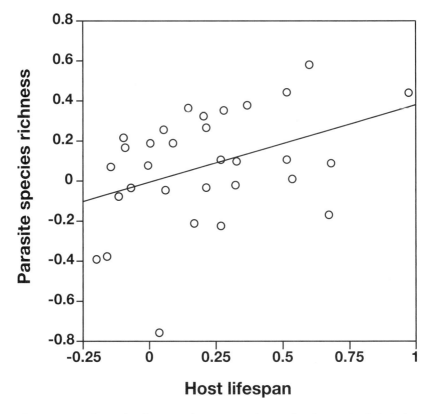

Figure 3.10. Relationship between the species richness of endoparasitic helminths and host lifespan, across species of North American freshwater fishes. Points are phylogenetically independent contrasts based on log-transformed data. (Data from Poulin and Morand 2000)

Epidemiological Models and Parasite Ecology

We now shift our focus from host features to properties of the parasite species that coexist in a host population. Using the concept of R_0 and complex multispecies epidemiological models, it is possible to explore what conditions promote the stability of an existing parasite assemblage in a host population and what conditions favor the invasion of additional species. The details of the models are available elsewhere (Dobson 1990; Dobson and Roberts 1994; Roberts and Dobson 1995); here, we look instead at two of the main predictions and how they fare when confronted with empirical data.

The models of Dobson and Roberts (1994) and Roberts and Dobson

(1995) begin by assuming that individuals from different parasite species are distributed independently among host individuals, and that interspecific competition is unimportant. The latter assumption requires some discussion here (the former is addressed later). There is ample evidence that different parasite species can compete in one form or another inside the same host individual (reviewed in Poulin 1998a). This is often seen in terms of reduced numbers, reduced fecundity or altered niche use by one species when a second one is also present. It can be argued, however, that parasites must first establish before they can compete. If the distributions of different parasite species among host individuals are independent, then establishment and persistence may supersede competitive interactions in determining the richness of parasite assemblages. In this scenario, the only way two or more parasite species would interact would be indirectly via their respective effects on the population dynamics of the host (i.e., if one virulent parasite species reduces host population density, there are fewer hosts left for other species). Under these conditions of no direct competition, some predictions emerge from the models (Dobson and Roberts 1994).

First, similarity in life-history traits among parasite species should allow more parasite species to coexist in the same host population. Specifically, the models predict that coexistence of many parasite species is more likely if they have similar values for life-history attributes involved in the determination of R_0, such as fecundity or lifespan (λ and μ, respectively, in the equations seen earlier), and associated life-history traits like body size. The models show that large differences between parasite species in life-history traits are likely to allow one or a few species to become dominant and exclude the others. We would thus expect to see increasing similarity in life-history traits among coexisting parasite species as species richness increases, among different host species. This is a difficult prediction to test given our extremely limited knowledge of parasite fecundity, lifespan, or life histories in general. It is interesting to note, however, that some of the most speciose assemblages of parasites known to coexist in the same host population often consist of several closely related parasite species that are highly likely to exhibit similar life-history traits. For instance, the rich parasite fauna of zebras includes about 40 nematode species, of which almost half are strongylids belonging to the same subfamily, Cyathostominae (Krecek et al. 1987). In horses, which also boast rich assemblages of intestinal helminth parasites, most species are strongylid nematodes belonging to only two subfamilies, and with a few genera accounting for most species (Bucknell et al. 1996). Similarly, the rich parasite assemblages of kangaroos and other macropodid marsupials are dominated by nematodes, again mainly strongylids and with some genera (e.g., *Cloacina*) represented by several species (Beveridge and Spratt 1996). Surely, the congeneric and confamilial parasite species in the preceding examples share many life-history features, and it is possible that this is a key factor allowing their coexistence. Of course, the evolutionary origins of closely related parasites in the same host population may involve a form of sympatric speciation, and these are apparently not frequent events (see Chapter 5).

Second, more parasite species can coexist in the same host population when their respective distributions among host individuals are highly aggregated. Aggregation refers to the clumped distribution of parasites among host individuals, with many host individuals harboring few parasites and most parasites being concentrated in few hosts (see Poulin 1998a). If the aggregated distributions of different parasite species are independent from one another, higher levels of aggregation will favor the coexistence of more parasite species because high numbers of different parasite species will be unlikely to co-occur in the same host. This creates a partitioning of the host population among parasite species. A similar phenomenon can occur in communities of free-living organisms inhabiting a patchy habitat (e.g., Ives 1991; Jaenike and James 1991). Modelling results suggest that if levels of aggregation are decreased, the maximum number of parasite species that can coexist will decrease, with species with the lowest levels of aggregation disappearing first (Dobson and Roberts 1994; Roberts and Dobson 1995). This prediction is also difficult to evaluate with empirical data. Almost universally, metazoan parasites are aggregated among host individuals in a population, whether they are from species-poor or species-rich assemblages (Dobson and Merenlender 1991; Shaw and Dobson 1995; Shaw et al. 1998). Whether levels of aggregation are generally higher in species-rich parasite assemblages than in species-poor ones has only been tested by Morand et al. (1999) on ectoparasite assemblages of marine fishes. Their results showed that as parasite species richness increased, so did levels of aggregation of individual parasites within species. Although further tests are required, it looks like parasite aggregation is indeed important in determining species coexistence and thus local parasite species richness within a host population.

At this point, a digression on parasite distributions is necessary. The above predictions rest on the assumption that the aggregated distributions of different parasite species among host individuals are independent from one another. Further, they assume that there is no direct competition between parasite species. If direct competition is added to the models, the equation for R_0 for each new parasite species becomes more complicated than those presented earlier, because the expression must now include a term for the mutual effects parasite populations have on one another (see Roberts and Dobson 1995). With competition present, the likelihood that a new parasite species invades an established assemblage is increased if there are predominantly negative correlations between the distributions of parasite species. In other words, if the probability of finding high numbers of individual parasites of more than one species in the same host individual is less than expected by chance alone, the chances of a new invading species are good. In natural assemblages of parasite species, negative pairwise correlations between numbers of individuals of different parasite species among host individuals are indeed common (e.g., Andersen and Valtonen 1990; Dezfuli et al. 2001). However, there are numerous striking exceptions. For instance, in lesser scaup ducks, 52 different species of intestinal helminths, mostly cestodes, were found from only 45 individual birds examined (Bush and Holmes 1986a). Of the 120 correlations of intensity of infection (number of individual parasites per host) that were computed between all pairs of the 16 most common species, all were positive and 31 of them were statistically significant. This is a rich assemblage of helminth species, and yet not a single species-pair covaries negatively in terms of the distributions of individual parasites. Positive covariance between the intensities of infection of different helminth species are also predominant in other systems (e.g., Haukisalmi and Henttonen 1993; Lotz and Font 1994; Holmstad and Skorping 1998). One explanation for this phenomenon could be that certain host individuals are susceptible to infection by all kinds of parasites, and thus accumulate all species of parasites because their defense mechanisms are inefficient. It may also be that the presence of one parasite species in a host facilitates the establishment of a second species, perhaps by depressing the host's immune system. In the cases of helminths with complex life cycles that are acquired by vertebrates via ingestion of infected prey, another possibility is that associations between parasite species in the prey are simply passed on to the assemblages in the vertebrate host (Lotz et al. 1995; Vickery and

Poulin 2002). Whatever the reasons for the predominance of positive covariances between the intensities of different parasite species in many natural systems, the implications for the predictions issued from epidemiological models are ambiguous. Either many existing assemblages of parasites cannot be invaded by new species because the aggregated distributions of existing species among host individuals are too strongly linked, or the importance of aggregation may have been overemphasized. Given the existence of rich assemblages where the distributions of all parasite species covary positively (e.g., Bush and Holmes 1986a), the latter possibility cannot be dismissed.

Another underlying assumption of many model predictions requires empirical confirmation. It is assumed that only the parameters that define R_0 for all the parasite species involved will determine how many, and which ones, can coexist in a host population. There is no consideration of whether an upper limit is placed on the number of parasite species that can share a host population set by other factors. If most natural parasite assemblages are saturated with species, for one reason or another, it will be impossible for an invading species to gain a foothold and spread, whatever its R_0 value. In communities of free-living organisms, a common test for species saturation consists of plotting local species richness against regional species richness for a range of localities and determining what function best fits the points (see Cornell and Lawton 1992; Srivastava 1999). In the context of parasite assemblages and their hierarchical structure (see Chapter 1), one could examine the relationship between maximum infracommunity richness and component community richness, that is, between the maximum number of parasite species found in one host individual and the number of parasite species in the host population, across different host populations or species. Two extreme scenarios are possible. The relationship may be linear, suggesting that the potential richness of infracommunities is generally proportional to the number of parasite species available in the component community, and that there is no saturation. Second, a curvilinear function would indicate that potential infracommunity richness becomes increasingly independent of component community richness as the latter increases, until the number of species reaches a plateau (i.e., saturation). Several caveats are associated with these simplified interpretations (Srivastava 1999), but still the shape of the function indicates whether species saturation is a possibility.

This approach has been applied to parasite assemblages on a few occasions. Among 64 component communities of intestinal helminths from different populations of eels in Great Britain, Kennedy and Guégan (1996) found a curvilinear relationship between infracommunity richness and component community richness and concluded that these parasite assemblages were saturated. However, not only is the relationship linear in American eels (Marcogliese 2001a), but alternative interpretations of the results of Kennedy and Guégan (1996) have been proposed, in which apparent saturation is only an artifact of other processes (Rohde 1998a). Kennedy and Guégan (1994) also found interspecific evidence for species saturation of helminth assemblages among British freshwater fish species. In contrast, in comparative studies across populations of different species of bird and mammal hosts, the relationship between maximum infracommunity richness and component community richness was linear (Poulin 1998a): there was no indication of species saturation (Figure 3.11). The same conclusion was also reached by Morand et al. (1999) in a comparative study of ectoparasite assemblages on marine fish species.

Another line of evidence suggests that species saturation is not a key feature of parasite communities, at least for helminths in vertebrates. The productivity of a parasite community, or the average standing biomass of parasites that can be supported by one host, might set an upper limit on parasite species richness. In fact, a linear function provides the best fit to the relationship between parasite species richness and parasite biomass per host, across parasite assemblages in different vertebrate host species (Figure 3.12); the result stands after correction for host body mass, which also correlates with parasite biomass. This suggests a certain complementarity among parasite species, with total biomass increasing with the addition of each new species. Given the evidence, it appears that species saturation is unlikely to be a major obstacle preventing the invasion of new species in parasite assemblages.

In summary, similar life-history traits and high aggregation levels are features of parasite species that can promote high species richness within host populations, according to epidemiological arguments. The available empirical evidence is still modest, but generally supports these predictions. It is also possible to extract from the expression for R_0 further predictions in relation to parasite diversity, such as which parasite features could promote parasite diversification via high speciation rates. These predictions will be addressed in Chapter 5.

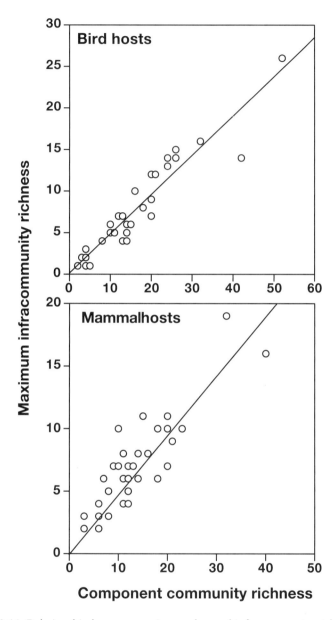

Figure 3.11. Relationship between maximum observed infracommunity richness and the species richness of the component community, across assemblages of intestinal helminths from 31 bird host species and 37 mammal species. For both sets of data, a curvilinear function did not provide a better fit to the data than the linear regressions shown. (Data from Poulin 1998a)

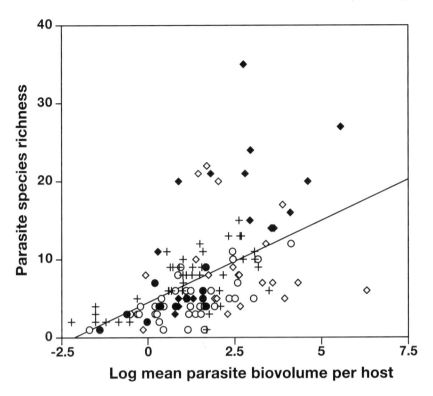

Figure 3.12. Relationship between parasite species richness and parasite biovolume (a surrogate of biomass), across 131 species of vertebrate hosts. The linear relationship shown ($y = 2.083x + 4.645$, $P = 0.0001$) holds even after correcting for two confounding variables: sampling effort and host body mass. Different host taxa are represented by different symbols: +, fish; filled circles, amphibians; open circles, reptiles; open diamonds, mammals; filled diamonds, birds. (Data from Poulin et al. 2003)

Conclusions

Epidemiological theory addresses the factors that determine the invasion success, spread, and persistence of parasites within and among host populations. It can also serve to make predictions about which host or parasite features may influence local parasite species richness. To date, most investigations have focused either on ectoparasites acquired via free-living infective stages or endoparasitic helminths acquired by ingestion. Epidemiological predictions for parasites transmitted by sexual contacts among hosts, or by vectors, would most likely be different, because invasion of a host population by such parasites is much less dependent on host density (see Morand 1993). The aim of this chapter was to present an epidemiological framework that can be accommodated to cover other types of parasites.

For ectoparasites and endoparasitic helminths, host population density, host lifespan, similarity in life history traits among parasite species, and parasite aggregation levels appear important predictors of parasite species richness, with varying levels of empirical support. These factors, however, are not the only ones that will affect parasite diversity. Several other processes, acting on much longer time scales, will also generate much variability in parasite diversity either among localities or among host populations or species; these will be examined in the following chapters. These other variables are not independent of the ones examined in this chapter. For instance, for most groups of vertebrates, local population density covaries positively with geographic range (Brown 1995; Rosenzweig 1995) and negatively with body size (Damuth 1987). Species with restricted geographic distributions or large body sizes occur at lower density than widespread species or small-bodied ones. Thus, in the next chapter, when the potential effects of host geographic range or body size on parasite species richness are discussed, remember that they are linked to other host features. The synergy of these myriad factors probably explains why, taken individually, none of them achieves great predictive power. Understanding the determinants of parasite species diversity requires a combination of epidemiological, ecological, and evolutionary perspectives.

4

Hosts as Drivers of Parasite Diversity

In a sample of host individuals taken from a natural population, the most striking feature of parasite infections is the high heterogeneity among hosts. Some host individuals harbor many species of parasites whereas others harbor few, and some hosts harbor much higher numbers of parasite individuals of a given species than other hosts. The efficiency of individual animals as samplers of locally available parasites varies greatly within a population. From the perspective of parasites, we could say that certain host individuals are easier to colonize than others. Many characteristics of host individuals have been identified as key determinants of the probability of acquiring parasites; these include simple factors such as age and sex as well as more complex ones such as behavior and immune status (see Wilson et al. 2002). Using the variability of these key features among individual hosts, we can predict which hosts are more likely to harbor many parasites.

A similar reasoning can be applied to the acquisition of parasite species over evolutionary time scales. Colonization of a host species by a parasite species is a process that may take several generations, and its success surely depends greatly on environmental conditions. Still, we might expect properties of the target host species to have a significant influence on its probability of being successfully colonized by a new parasite species. In the previous chapter, we saw that host population density and host lifespan can affect the diversity of parasites harbored by a host population or species. In this chapter, we will explore the role of other host features in determining how parasite diversity is partitioned among host species. The chapter begins with an introduction to host–parasite coevolution and details the importance of examining parasite species richness in a phylogenetic context. This is followed by a discussion of the application of island biogeography theory and related ideas to the study of parasite species richness; these have provided much of the contextual background for empirical studies to date. Finally, we summarize the available evidence in support of a role for some host features in determining parasite species richness.

Host–Parasite Coevolution: Association by Descent and Colonization
Since parasites are by definition associated with their hosts, their evolutionary history can be mapped onto host phylogeny (Page 2003). Throughout the phylogenetic history of a group of host species, parasite species will be acquired or lost over time; the different rates of parasite acquisition and loss produced the current distribution of parasite species richness among living host species. The phylogeny of the hosts represents the history of the living habitats available to parasites, and we must consider how these habitats accumulate parasite species.

As discussed later, several ecological processes can influence the probabilities of parasite extinction, colonization, and even speciation within a given host population (parasite component community) or host species (parasite fauna). From an evolutionary perspective, there are three general explanations for the origins and presence of any parasite species in a parasite fauna (Page 1994; Hoberg et al. 1997; Paterson and Gray 1997; Vickery and Poulin 1998). First, a parasite species may have been inherited by the host species from its ancestor: the parasite species of an ancestral host species may be passed down to daughter host species during host speciation (Figure 4.1). Such parasite species will be shared with closely related host species. In fact, the parasites on sister host species may have cospeciated on their own at about the same time as the hosts diverged and may now be different parasite species, but still sibling species inherited from the same ancestor. Inherited parasite species like these are not the product of ecological processes; that is, they are not acquired because of divergence in ecological features of their hosts. They are evolutionary baggage, the product of a form of phylogenetic inertia, and, although they contribute to the richness of the parasite fauna of a host species, their presence may not reflect recently acquired host characteristics.

Second, a parasite species may have colonized the host species, jumping ship from another, sympatric host species (Figure 4.1). In principle, host

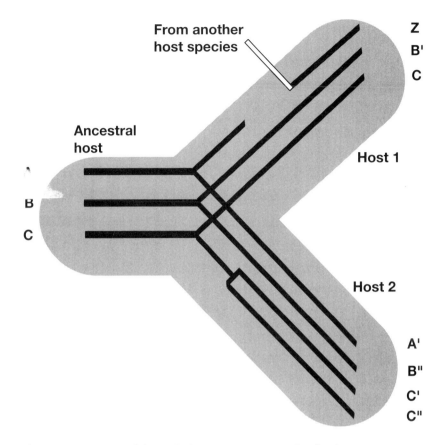

Figure 4.1. Summary of the evolutionary events accounting for the present richness of different parasite assemblages. Host species 1 and 2 are sister taxa issued from a common ancestral host species, as indicated by the divergence of the shaded area. They have each inherited certain parasites from their ancestor, such as parasite lineage B (called B′ and B″ in the daughter host species because the parasites may have speciated as well). New parasite species may also be acquired via intrahost speciation (e.g., parasite lineage C in host 2) or by colonization from sympatric host species (parasite Z in host 1). Parasite lineages may also go extinct (e.g., parasite A in host 1).

switching is possible provided that the new host species is ecologically, immunologically, and physiologically similar to the original host. Host switching can be an important contributor to parasite species richness in any given host species. For instance, among North American freshwater fish species, generalist parasite species that exploit several different host species after repeated host switches are the main component of the parasite faunas of many host species (Poulin 1997a). After the switch, parasites may also speciate on their new host, leading to a completely new host–parasite association; some human parasites appear to have been acquired in this way (Combes 1990; Waters et al. 1991).

Third, a parasite species may be the outcome of an intrahost speciation event: an ancestral parasite species giving rise to one or more daughter species all within the same host species, without host speciation (Figure 4.1). This phenomenon may not be common, but it could account for the presence of many congeneric parasite species in certain host species (Kennedy and Bush 1992; Poulin 1999a; see Chapter 5).

There are also two ways in which parasite species may be lost from component communities or parasite faunas (Paterson and Gray 1997; Vickery and Poulin 1998). First, a parasite species can miss the boat during a host speciation event, for instance when the founder host population giving rise to a new daughter host species happens to be parasite-free. Because most metazoan parasites are aggregated among host individuals in a population (Shaw and Dobson 1995; Shaw et al. 1998), most host individuals harbor few or no parasites. It is therefore conceivable that parasites of a given species can be absent from a founder host population, or present in such small numbers that they cannot persist. Second, inherited parasite species can go extinct well after the host speciation event (Figure 4.1), as can parasite species acquired via host switching. Parasite extinction may result from several factors. For example, the host may become resistant, the parasite may be excluded by colonizing parasite species, other host species necessary for the completion of the parasite's life cycle may disappear, or environmental conditions may change and become inhospitable for the parasite's free-living stages.

The frequency of these events over evolutionary time is no doubt relatively low, but they are probably not so rare that Figure 4.1—in which two sister host species display quite different parasite faunas—becomes merely a caricature. After a speciation event, newly formed host lineages may diverge in body mass, diet breadth, geographic range, or other ecological trait. The likelihood of losing or acquiring parasite species by a host species (i.e., by a parasite fauna) over evolutionary time is most probably related to the ecological characteristics of the host species. The rest of this chapter will explore the influence of these ecological traits in shaping parasite diversity.

Before moving on to discuss the role of host ecological traits, however, two important issues must be addressed. The first issue concerns the relative importance of host switching versus cospeciation. In other words, how many of a host species' parasites are the product of colonization rather than legacies from an ancestral host? To reconstruct the coevolutionary history of parasite species with their hosts, we must begin by comparing their phylogenetic trees and assessing their degree of congruence. If the phylogeny of a group of related parasite species is an exact mirror image of that of their host species (i.e., if the parasite phylogeny is perfectly congruent with the host phylogeny), then we have a case in which all parasite species on extant hosts are the product of cospeciation events (Figure 4.2). If, on the other hand, there is absolutely no congruence between the phylogenies of hosts and parasites, host switching has been rampant. There are potential problems, however. False congruence can arise from a sequential colonization of host species by parasites which coincidentally reflects the pattern of host radiation (Brooks and McLennan 1991). There may also be false incongruence between host and parasite trees (Figure 4.2), resulting from intrahost speciation by parasite lineages, as well as from parasites "missing the boat" or becoming extinct (Page 1993; Paterson and Gray 1997). Empirical evidence suggests that parasites often miss the boat—they are often absent from founder host populations that can be the prelude to host speciation events (Paterson et al. 1999, 2003). It is therefore not straightforward to determine the relative importance of host switching based on the congruence of host and parasite phylogenies.

At present, two main analytical methods are used to assess the degree of congruence between host and parasite phylogenetic trees. One method, Brooks' parsimony analysis (BPA), treats parasites as host character states and treats parasite phylogeny as a character state tree; it then compares the most parsimonious host cladogram derived from the parasite data with the original host phylogeny obtained from independent characters (Brooks and McLennan 1991). Incongruence is taken as evidence for one or more host switching events. This maximizes, perhaps even inflates, the number of host

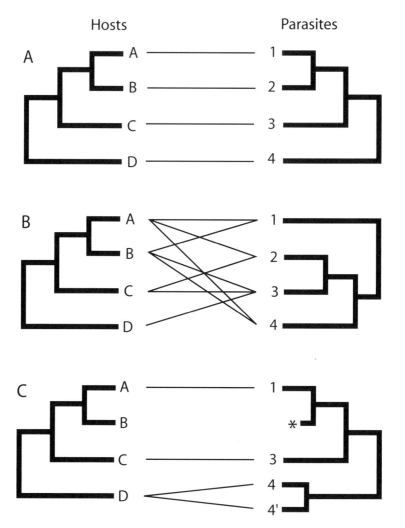

Figure 4.2. Phylogenetic trees of hypothetical host and parasite species. The thin lines indicate host–parasite associations. (A) Perfect congruence between the trees, indicating strict host–parasite cospeciation. (B) Incongruence resulting from frequent host switching by the parasites. (C) False incongruence resulting from parasite extinction (*) and intrahost parasite speciation.

switches detected for any host–parasite association. The other method, reconciliation analysis in its various versions, contrasts independently derived host and parasite phylogenies and determines whether their congruence is greater than expected by chance (Page 1990, 1993, 1994). The reconciliation approach explains incongruence in terms of intrahost parasite speciation or parasite extinction, without postulating any host switches; in other words, this method maximizes the frequency of cospeciation events. Recent versions of reconciliation analysis allow for host switching (Page 1994), and further developments of the method even allow the differential weighting of all events (cospeciation, host switching, intrahost speciation, extinction) rather than simply maximizing the frequency of cospeciation events (Paterson and Banks 2001). This is an important step toward a unified analytical method without *a priori* biases in favor of one or another kind of evolutionary origin for parasite species. Still, the debate over the advantages and pitfalls of both methods continues (see Dowling 2002; Dowling et al. 2003; Page 2003; Siddall and Perkins 2003).

Analyses using either of these two methods have revealed some interesting trends, but no universal scenario for host–parasite coevolution (Page 2003). Many studies on ectoparasites of vertebrates (e.g., lice, mites) have revealed relatively tight patterns of cospeciation between host and parasites (Hafner and Nadler 1988, 1990; Paterson et al. 1993; Paterson and Poulin 1999; Dabert et al. 2001). This is not the case for other systems involving lice on birds or mammals (Barker 1991; Clayton et al. 1996; Johnson et al. 2002a; Taylor and Purvis 2003) and many associations between helminths and vertebrates (e.g., Hoberg 1992, 1995; Brooks and McLennan 1993b; Hoberg et al. 1997; Skerikova et al. 2001; Cribb et al. 2002a; Desdevises et al. 2002). Similarly, patterns of relatively strict cospeciation are also rare in plant–phytophagous insects systems (Farrell et al. 1992; Price 2002). Coevolutionary trajectories therefore vary among host–parasite systems, and we cannot simply assume that cospeciation or host switching has been the predominant mode of evolution.

On a large taxonomic scale, however, it is possible to test a simple prediction derived from the assumption that cospeciation has been the most frequent mode of coevolution between parasites and their hosts. If gene flow is interrupted between two allopatric populations of hosts such that they and their parasites speciate at roughly the same time, and if this process of cospeciation is repeated several times in the daughter species, we would ex-

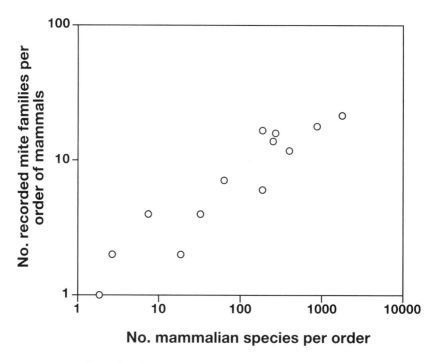

Figure 4.3. Relationship between the number of known mite families (parasitic or not, but obligate associates of mammals) per host taxon (i.e., mammalian orders), and the species richness of the host taxon. (Data from Walter and Proctor 1999)

pect n host species in a particular taxon harboring n parasite species of a given taxon, each species being highly host-specific. There is evidence that this kind of codiversification has happened among malaria parasites of birds (genera *Plasmodium* and *Haemoproteus*), where the number of parasite lineages almost equals the number of host species sampled (Ricklefs and Fallon 2002). As a consequence of this process, the species richness of different parasite taxa should be proportional to that of the host taxa they exploit: the more speciose host taxa should harbor more parasite species than their less speciose sister taxa. High diversification of parasites in speciose host taxa appears to hold for certain groups of parasitoid wasps and their insect hosts and for phytophagous insects and their plant hosts (e.g., Price 1980; Mardulyn and Whitfield 1999). Among true parasites, this relationship holds for mites and their mammal hosts, with the most speciose mammalian orders harboring the highest diversity of mite families (Figure

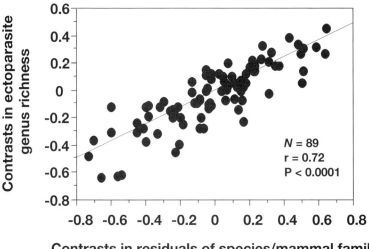

Figure 4.4. Relationship between the number of genera of ectoparasitic arthropods per host taxon (i.e., mammalian families) and the species richness of the host taxon. Points are phylogenetically independent contrasts that have been corrected for a confounding variable, mean host body mass, by using residuals from a regression. (Data from Kim 1985)

4.3). The pattern holds well when all ectoparasitic arthropods are examined: the most speciose mammalian families harbor the greatest diversity of ectoparasites (Figure 4.4). Not surprisingly, the pattern does not hold across all groups. For instance, the more diverse bee families are not hosts to more taxa of mites than the less diverse bee families (Walter and Proctor 1999). The large-scale distribution of trematode species among vertebrate groups is also not compatible with strict cospeciation. Mammals account for only a small proportion of known vertebrate species, yet they host nearly one-quarter of known trematode species; amphibians and reptiles, on the other hand, represent a quarter of known vertebrates, but are hosts to less than 10% of known trematodes (Figure 4.5). Parasite origins other than inheritance from a common ancestor, such as intrahost speciation or host switching, must therefore often be important contributors to parasite species richness.

Before discussing the role of host ecological traits in the diversification of parasite faunas, we must address the need to control for phylogenetic effects in comparative tests. The phylogenetic inertia that leads to sister host taxa

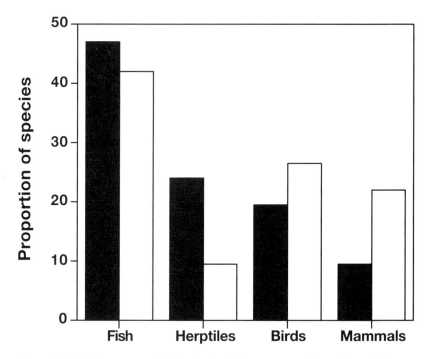

Figure 4.5. Relative trematode diversity in different groups of vertebrates. Shown are the proportions of the total number of known vertebrate species belonging to four main taxonomic groups (black bars), with herptiles including reptiles and amphibians, and the proportions of known trematode species parasitizing each of the host groups (open bars). Vertebrate diversity data are from Barnes (1998) and trematode data from Gibson and Bray (1994). (From Poulin and Morand 2000)

inheriting parasite species from their common ancestor means that sister host species share parasite species (or parasite lineages if the parasites have cospeciated) simply because they are related. They may have diverged in ecological characteristics and acquired new parasite species since they were separated, but the baggage of shared parasites they carry can mask any effect their ecological traits have had on the diversification of their parasite faunas. Consider two sister host species that have diverged in body size, with species A having a body mass of 30 g and species B having a mass of 60 g. If species A harbors 23 parasite species and species B harbors 26 species, we might be tempted to conclude that a twofold difference in body mass has almost no influence on parasite species richness. However, if each host species has inherited 20 parasite species from their common ancestor, they would have acquired three and six new parasite species, respectively, since diverging. Without taking into account the similarities between the two host species that are due to their common phylogenetic origins, we could have erroneously dismissed host body size as a determinant of parasite species richness. The same argument can be made for any other host trait, such as geographic range or diet, and phylogenetic influences on all these traits must be accounted for when comparing different species (Brooks and McLennan 1991; Harvey and Pagel 1991). Another way to put it is that related host species are not independent in a statistical sense. In the context of parasite species richness, the most commonly employed method to account for non-independence and confounding phylogenetic effects has been the phylogenetically independent contrast method (Felsenstein 1985). The method has survived the test of time and been shown to be robust to a range of background noise (e.g., Diaz-Uriarte and Garland 1998; Oakley and Cunningham 2000). It consists of deriving contrasts, or differences, in the values for different traits between sister species in a phylogeny, and using these sets of contrasts in standard statistical analyses. For instance, using the sister host species A and B above, we would obtain a contrast in body mass of +30 g (60 minus 30) and a contrast in parasite species richness of +3 species (26 minus 23). There is thus a positive association between host body mass and parasite species richness in this hypothetical example; obtaining several such sets of contrasts from real host species would allow one to perform a rigorous test of the effect of host mass. The different results often obtained before and after controlling for phylogeny (e.g., Poulin 1995c) highlight the potential importance of this confounding variable. It should always be taken into account, in addition to sampling effort (see Chapter 2), in comparative analyses of parasite species diversity used to test the ideas presented in this and other chapters.

Hosts as Islands

If parasite faunas (or parasite component communities) become richer via intrahost parasite speciation or when new parasite species colonize the host species, and if they lose species through local or large-scale parasite extinction, what are the features of host species that facilitate intrahost parasite speciation, host switching, or conversely parasite extinction? The greatest source of inspiration for ecologists interested in parasite species diversity over the past 20 years has been MacArthur and Wilson's (1967) theory of island biogeography. The theory of island biogeography postulates that the equilibrium number of species on an island reflects the balance between the rate at which new species colonize the island and the rate at which species go extinct on the island. These rates are influenced by various features of islands, in particular their size, their age, or their distance from the mainland. Larger islands provide bigger targets for colonizing species arriving by air or on drifting plant material, more space, and a greater variability of habitats; they should thus receive more species that can reach larger population sizes, allowing them to escape local extinction. New islands, for instance those recently produced by volcanic activity, begin their existence free of animal or plant inhabitants and accumulate them over time; we might thus expect that, all else being equal, species richness will increase asymptotically with island age. Finally, distance from the mainland will also affect rates of species colonization since continents are the main source of colonizers; all else being equal, remote oceanic islands should therefore be inhabited by fewer species than those lying along continental land masses.

Although the analogy shows some weaknesses, hosts can be viewed as islands for parasite colonization at several hierarchical levels (Kuris et al. 1980). An individual host animal is born parasite-free and accumulates parasites over its lifetime depending on its exposure to infective stages; thus, both its age and its proximity to sources of infection affect the number of parasite species it acquires. At higher scales, the number of parasite species exploiting a host population (component community level) or a host species (parasite fauna level) should be the product of the rate at which new parasite species colonize the host or arise by intrahost speciation and the rate at which they go extinct. There is one key difference between hosts and true islands, however: whereas new islands are free of animal and plant species, new host species inherit some parasite species from their ancestors (Figure 4.1). Thus rates of colonization and extinction will cause related host species to have different parasite species richness, but they will most likely have common components in their parasite faunas from the start. This serves to emphasize again why it is crucial to correct for phylogenetically inherited parasite species in comparative analyses across host species.

By drawing parallels with island biogeography, several ecological features of the host are predicted to be important modulators of the rates of parasite acquisition and parasite loss. For instance, host species with large geographic ranges will have a distribution that overlaps with the ranges of several other host species, and they will therefore have more chance of acquiring new parasite species via colonization. This is similar to islands that are close to the mainland having higher probabilities of being colonized by mainland species. We would thus expect that, all else being equal, there should be a positive, interspecific relationship between parasite species richness and host geographic range (Figure 4.6). By analogy to island age, host lifespan can be seen as a key factor affecting rates of parasite colonization. There is some, not strong, evidence that longer-lived host species may harbor more species than short-lived ones (see Chapter 3). However, in the context of island biogeography, the age of the host population or species may be more relevant than the lifespan of the host individual: after their establishment in a new area, host populations may accumulate parasite species until they attain some maximum (equilibrium) species richness. Finally, host body size is also often invoked as a determinant of the rates of parasite colonization (or intrahost speciation) and extinction based on island biogeography arguments. Like larger islands, larger-bodied host species provide more space and other resources for parasites, and possibly a greater diversity of microhabitats, and may thus be able to support richer faunas. Larger-bodied host species are bigger targets for parasite colonization because they ingest food (and thus food-borne parasite infective stages) at higher rates, and because they have a greater surface area for contact and attachment by parasite infective stages. The assumptions on which the link between host body size and parasite species richness is based are basically the same that support species-area relationships in free-living assemblages (see Connor and McCoy 1979; Rosenzweig 1995; Gaston and Blackburn 2000). One could argue that the equivalent of island size to parasite species richness is not host body size but rather host biomass. A single elephant per square kilometer will not be capable of supporting as many parasite species as a hugely abundant rodent species with much smaller body mass but much higher biomass per unit area. As biomass is the product of body mass and density, and given the established role of host population density in determining parasite species richness (see Chapter 3), we will here concentrate on host body size.

The predictions made by the theory of island biogeography were quantitative in nature (MacArthur and Wilson 1967). As a rule, predictions made about parasite species richness are qualitative rather than quantitative. In

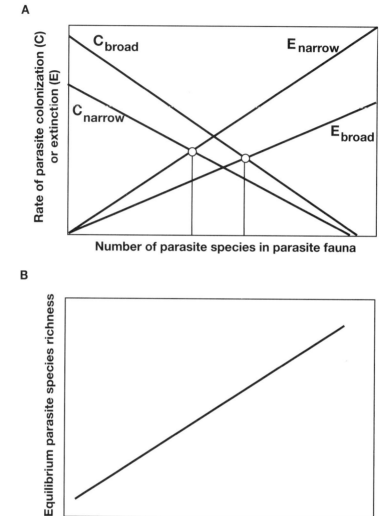

Figure 4.6. Use of biogeography theory to make predictions regarding parasite species richness. (A) The equilibrium parasite species richness in a host population or species reflects the balance between the rates of parasite colonization and extinction (i.e., the intersection of the lines in the figure), as shown here for hosts with either a narrow or a broad geographic range. Colonization rates are assumed to be higher, and extinction rates lower, in host species with broad geographic ranges. (B) Predicted relationship between host geographic range and parasite species richness, from (A).

other words, a general increase in parasite species richness is expected with increasing host body size, but there have been no attempts to predict how many more parasite species should be expected when host body size is, say, doubled. Although species–area relationships for free-living organisms assume reasonably consistent functions with a limited range of parameter values (see Rosenzweig 1995; He and Legendre 1996; Gaston and Blackburn 2000), similar relationships between host body size (or any other host trait) and parasite species richness are reported merely as correlations in most studies; at this point quantitative relationships are unknown.

Host Traits and Parasite Species Richness

In addition to host geographic range, the age of host populations, host body size, and the variables discussed in Chapter 3, several other host ecological traits have been shown to influence parasite species richness in certain systems (reviewed in Poulin 1997b, 1998a; Morand 2000). All of these traits affect the likelihood that parasite species will colonize and/or persist in new host species, for instance, by influencing either the types of parasites that the host will encounter, how easy it will be for them to be transmitted among host individuals, or the permanency of the host as a habitat. The available evidence for a role of these variables in determining parasite diversity is summarized in Table 4.1, and they will all be discussed below. The usual approach in seeking to explain parasite species richness with these host traits is to search for a pattern across several different host species that matches, at least qualitatively, an *a priori* prediction. This approach follows the recent development of macroecology and similar attempts to explain large-scale patterns in species diversity (e.g., Huston 1992; Ricklefs and Schluter 1993; Brown 1995; Rosenzweig 1995; Maurer 1999; Gaston and Blackburn 2000). This approach does not allow the experimental demonstration of causality, but rather the inference that one or more factor are influencing parasite species richness from independent demonstrations that they are associated with parasite richness in many assemblages. This is perhaps not the strongest line of evidence in science, but it has been championed as the best way of uncovering general laws in ecology (Lawton 1999).

Before discussing the various host features that may be linked with parasite diversity, we present one compelling piece of evidence that parasite species are *not* distributed at random among host species, and that there must be something special about certain host species that makes them magnets for

Table 4.1

Summary of the empirical evidence linking parasite species richness with some of the best-studied host characteristics, obtained from comparative studies

Host trait	Host group	Parasites	Correlation?	Study
Body size				
	Teleosts	Helminths	Yes	Price and Clancy 1983
	Teleosts	Helminths	Yes[a,b]	Bell and Burt 1991
	Teleosts	Monogeneans	Yes	Guégan et al. 1992
	Teleosts	Helminths	Yes[a,b]	Poulin 1995c
	Teleosts	Ectoparasites	No[a,b]	Poulin 1995c
	Teleosts	Ectoparasites	Yes	Rohde et al. 1995
	Teleosts	Helminths	Yes[a,b]	Gregory et al. 1996
	Teleosts	Ectoparasites	No[a,b]	Poulin and Rohde 1997
	Teleosts	Helminths	No[a,b]	Sasal et al. 1997
	Teleosts	Ectoparasites	Yes[a,b]	Morand et al. 1999
	Teleosts	Helminths	No[a,b]	Morand et al. 2000
	Birds	Helminths	No[b]	Gregory 1990
	Birds	Helminths	Yes[a,b]	Gregory et al. 1991
	Birds	Helminths	No[a,b]	Poulin 1995c
	Birds	Helminths	No[a,b]	Gregory et al. 1996
	Birds	Ectoparasites	No[a,b]	Clayton and Walther 2001
	Mammals	Helminths	No[a,b]	Poulin 1995c
	Mammals	Helminths	No[b]	Watve and Sukumar 1995
	Mammals	Helminths	No[a,b]	Gregory et al. 1996
	Mammals	Helminths	No[a,b]	Nunn et al. 2003a
	Mammals	Fleas	Yes[b]	Krasnov et al. 1997
	Mammals	Helminths	No[a,b]	Morand and Poulin 1998
	Mammals	Nematodes	Yes[a,b]	Arneberg 2002
Geographic range				
	Teleosts	Helminths	Yes	Price and Clancy 1983
	Teleosts	Helminths	No[b]	Gregory 1990
	Teleosts	Helminths	Yes[a,b]	Bell and Burt 1991
	Teleosts	Helminths	Yes	Aho and Bush 1993
	Teleosts	Helminths	Yes	Guégan and Kennedy 1993
	Teleosts	Monogeneans	Yes[a,b]	Simkova et al. 2001
	Herptiles	Helminths	Yes	Aho 1990
	Birds	Helminths	Yes[b]	Gregory 1990
	Birds	Helminths	No[a,b]	Gregory et al. 1991
	Birds	Lice	No[a,b]	Clayton and Walther 2001
	Mammals	Mites	Yes	Dritschilo et al. 1975
	Mammals	Mites	No[b]	Kuris and Blaustein 1977
	Mammals	Helminths	No[b]	Watve and Sukumar 1995
	Mammals	Helminths	Yes[a,b]	Feliu et al. 1997
	Mammals	Fleas	No[b]	Krasnov et al. 1997

Table 4.1
Continued

Host trait	Host group	Parasites	Correlation?	Study
Diet				
	Teleosts	Helminths	Yes[a,b]	Bell and Burt 1991
	Teleosts	Helminths	Yes	Guégan and Kennedy 1993
	Teleosts	Helminths	No[a,b]	Poulin 1995c
	Teleosts	Helminths	No[a,b]	Sasal et al. 1997
	Teleosts	Helminths	Yes[a,b]	Morand et al. 2000
	Birds	Helminths	No[a,b]	Gregory et al. 1991
	Birds	Helminths	No[a,b]	Poulin 1995c
	Mammals	Helminths	No[a,b]	Poulin 1995c
	Mammals	Helminths	No[b]	Watve and Sukumar 1995

[a] Corrected for phylogenetic influences.
[b] Corrected for unequal sampling effort among host species.

parasites. Ecologists have long been aware of the positive relationship between the numbers of species of different taxonomic groups among different localities and have proposed theoretical explanations for these relationships (Gaston 1996). Similarly, we can observe significant covariation in the species richness of different groups of parasites among different host species. Among vertebrate host species, the species richnesses of trematodes, cestodes, nematodes, and acanthocephalans tend to covary positively (Figures 4.7 and 4.8). In fact, of 30 pairwise correlations that we computed between the species richness of two parasite groups across host species in the different higher groups of vertebrates, 21 significant positive relationships were found (Figure 4.8). This high frequency of covariation in parasite species richness indicates that if a host species accumulates, say, trematode species at a high rate over time, it will also accumulate many other types of parasite species. In contrast, a host species with relatively few trematodes will tend to harbor a poor fauna of other parasites. Variation among host species in the features that we discuss next would be a logical explanation for these patterns.

Host Geographic Range
On small spatial scales and over short periods of time, overlap among host species can facilitate the transfer of parasite species among host species and

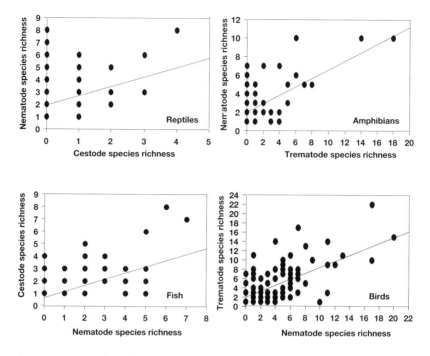

Figure 4.7. Examples of positive relationships between the species richness of one parasite group and that of another, across host species in different vertebrate taxa. (Data from Bush et al. 1990; Poulin 1995c)

increase the species richness of the parasite assemblages harbored by individual host species. For instance, the species richness of gastrointestinal parasites of African bovids increases with the number of other bovid species occurring in sympatry (Ezenwa 2003). On larger spatial scales, host species with broad geographic ranges will overlap with the ranges of many other host species from which they can acquire new parasites over long time scales. They will also encounter a broad range of environmental conditions that should result in their exposure to a wide spectrum of parasite species. Thus we expect host species with broad geographic ranges to have richer parasite faunas than host species with restricted ranges.

Several early tests of this prediction found empirical support, but failed to control for potential phylogenetic or sampling effects. For example, Dritschilo et al. (1975) found that host geographic range was a good predictor of the species richness of ectoparasitic mites on rodents, but this re-

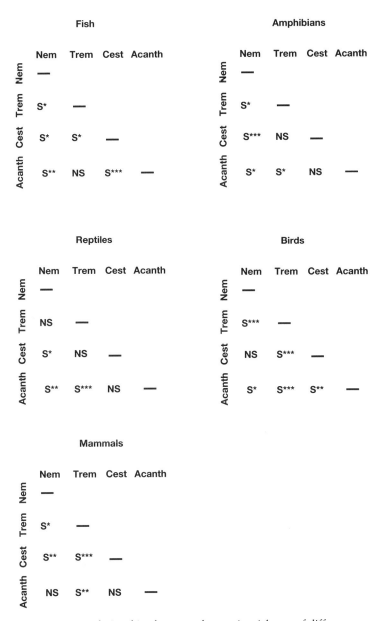

Figure 4.8. Pairwise relationships between the species richness of different groups of helminth parasites (nematodes, trematodes, cestodes, and acanthocephalans) among species of vertebrate hosts (NS = nonsignificant relationship, S = significant positive relationship, * $P < 0.05$, ** $P < 0.01$, *** $P < 0.001$). (Data from Bush et al. 1990; Poulin 1995c)

lationship was later found to be an artifact of unequal sampling effort (Kuris and Blaustein 1977). Many of the studies that followed also reported generally strong positive correlations between host geographic range and parasite species richness in a range of host taxa, but without correcting for both sampling effort and phylogenetic influences (Price and Clancy 1983; Aho 1990; Bell and Burt 1991; Aho and Bush 1993). There were some exceptions, where no significant relationship was found between host geographic range and parasite species richness (Guégan and Kennedy 1993; Watve and Sukumar 1995; Krasnov et al. 1997).

Among the few studies that took into account both host phylogeny and sampling effort as confounding variables, there is no general agreement. Gregory (1990) found a robust association between helminth parasite species richness and the geographic range of their waterfowl hosts, after controlling for all possible confounding variables. Among a larger sample of bird species from the former Soviet Union, Gregory et al. (1991) found no relationship between host geographic range and the richness of their helminth parasite fauna. Feliu et al. (1997) found that helminth species richness increased with increasing host geographic range among rodent species occurring on the Iberian Peninsula (see also Goüy de Bellocq et al. 2002). Simkova et al. (2001) found that host geographic range covaried with parasite species richness among European freshwater fish species, but proposed an indirect mechanism to account for the relationship. Finally, among Peruvian birds, Clayton and Walther (2001) found no association between geographic range and the species richness of chewing lice. There is thus support for a role of host geographic range, but it is not strong or applicable to all systems (Table 4.1). Perhaps this is not surprising. Host geographic range correlates tightly with sampling effort: the larger the range of an animal species, the more likely it is to encompass many universities and other research institutes, and the more accessible it is to biologists carrying out parasite surveys. Thus correcting for sampling effort can amount to removing the true effect of geographic range, making it difficult to detect its action (Guégan and Kennedy 1996).

Host individuals and populations are not distributed evenly across the entire geographic range of the host species. Hosts occur in distinct populations, more or less interconnected with each other. At the level of the parasite component community (the assemblage of parasite species co-occurring in the same host population), the parallel with island biogeography becomes clearer. Other conspecific host populations act as the main sources of colonizing parasite species, and their proximity becomes a key factor, just as proximity from the mainland matters for species richness on oceanic islands. Indeed, species richness of helminth parasites in lake populations of fish hosts is often mainly determined by the degree of geographic isolation of the lakes from one another (Poulin and Morand 1999). Distances among lakes will matter less for certain types of parasites, for example, those disseminated from lake to lake by bird definitive hosts (Esch et al. 1988), but generally speaking the presence of adjacent populations facilitates the colonization of a host population by many parasite species.

Host Body Size
Larger-bodied host species should provide more space, more nutrients, and possibly a wider variety of niches for parasitic organisms. They also feed at higher rates, a feature that exposes them to more food-borne parasite infective stages per unit time, and they tend to live longer than small-bodied hosts, providing a more permanent habitat for parasites. We might thus expect an increase in parasite species richness with increasing host body size. And indeed, several investigators have assessed the influence of host body size, measured as mass or length, on parasite species richness. Practically all studies that did not control for both unequal sampling effort and phylogenetic influences reported positive associations between host size and parasite richness (Price and Clancy 1983; Bell and Burt 1991; Guégan et al. 1992; Rohde et al. 1995; Watve and Sukumar 1995; Krasnov et al. 1997). These results suggest that at some level there must exist an association between the size of the host and the number of parasite species they accumulate.

As with host geographic range, the more conservative analyses that correct for both host phylogeny and parasite sampling effort provide a mixed bag of results (Table 4.1). Using essentially the same database, one study concluded that host body size was generally not an important determinant of helminth species richness in vertebrates (Poulin 1995c), whereas another study found that host body size strongly correlates with parasite species richness (Gregory et al. 1996). Two studies of ectoparasite species richness on marine fish species also generated contrasting results despite using practically the same data (Poulin and Rohde 1997; Morand et al. 1999); the likely reason is the fact that the two studies took different sets of confounding variables into account. Results from other studies, all using independent datasets, are just as inconsistent (Gregory et al. 1991; Sasal et al. 1997; Morand and Poulin 1998; Clayton and Walther 2001; Arneberg 2002). Intuitively, one would expect host body size to be a good predictor of how many species of parasites the host accommodates. With the empirical evidence so fickle and apparently so dependent on the details of the statistical analysis used, however, the safest conclusion is that host body size can be an important factor, but its role is far from overriding or universal.

Age of the Host Population

A new island begins its existence empty, free of life. A new host population, or a new host species, will usually include one or a few parasite species that were present in the founder subpopulation. Over time, through colonization by parasite species from adjacent host populations (whether conspecifics or not) and by intrahost parasite speciation, the number of parasite species exploiting a host population will increase toward some equilibrium number where it will stabilize. We should therefore expect that the age of host populations should determine how many parasite species they harbor. Clearly, beyond a certain age, and all else being equal, all populations will have reached their equilibrium number of parasite species, and host population age will no longer correlate with parasite species richness; the importance of host population age will only apply to populations in the early stages of their existence. Similar arguments have been made for the species richness of phytophagous insects on trees (Southwood and Kennedy 1983; Kennedy and Southwood 1984).

Although plausible, this idea is difficult to test because of the time scale involved. At its largest scale, the hypothesis that host population age is a good predictor of parasite species richness could be extended to predict that phylogenetically older host lineages harbor more parasite species than younger ones. Bush et al. (1990) have shown how difficult it is to evaluate this prediction, especially if working at high taxonomic levels as they did in comparing helminth parasite species richness among broad groups of vertebrates.

A better way of testing this hypothesis involves the use of historical data on actual populations rather than phylogenetic lineages. Guégan and Kennedy (1993) proposed that the richness of the helminth fauna of British freshwater fish species was influenced by the time since their arrival in Britain from continental Europe. Because of founder effects, introduced fish species usually bring with them only few of the parasite species that exploit them in their area of origin, and they accumulate new ones after their establishment in a new locality. Guégan and Kennedy (1993) argue that the relative arrival times of fish species in Britain provides a better explanation of the variation in the richness of their parasite faunas than their current geographic range in Britain or other ecological features. The acquisition of parasite species by introduced fish occurs via host-switching (i.e., colonization of the introduced fish by parasites of native fishes). Thus the species richness of introduced fish is proportional to the duration of their exposure to the local pool of parasite species. A similar scenario has unfolded in the case of our species: the number of parasite species we share with various domestic animals is proportional to the time since their domestication (Southwood 1987).

Working on a much smaller time scale, Ebert et al. (2001) have shown that host population age is the best predictor (explaining about 50% of the variance) of the richness of the endoparasite and epibiont fauna of rock pool populations of the crustacean *Daphnia magna*. Their study indicated that even more than 15 years after its establishment in a rock pool, a *Daphnia* population still accumulated new endoparasite species, whereas the number of epibiont species per *Daphnia* population saturated after just a few years (Figure 4.9). This is solid evidence that host population age is a key determinant of parasite species richness, at least for certain types of parasites or symbionts. Ebert et al.'s (2001) system did not include metazoan parasites and focused on shorter time periods than those relevant to most vertebrate host populations and their metazoan parasites, but it provides a microcosm of what would happen with longer-lived hosts and parasites with longer generation times and slower dispersal rates. In fact, host species with short generation times like *Daphnia* will provide the best opportunities for further tests of this relationship.

Host Habitat

Differences in habitat characteristics could also explain differences in parasite species richness between host populations or species. Isolated habitats can limit the exposure of hosts to new parasites and prevent the acquisition of novel parasite species from other host species if these are too far away (e.g., Marcogliese and Cone 1991a). More generally, certain environmental factors can influence the probability that a given parasite species

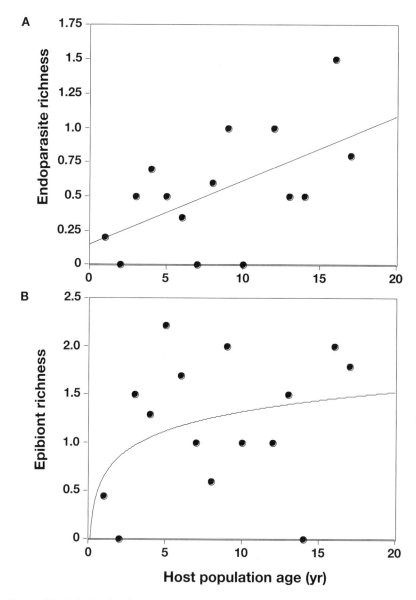

Figure 4.9. Relationship between the species richness of endoparasites (A) and epibionts (B) found in rock pool populations of the crustacean host *Daphnia magna* and the age of the populations. Each point represents the average richness of several pools of the same age, from a total of 86 pools; the oldest age category includes populations ≥17 years. The lines follow the analyses in the original study. (Data from Ebert et al. 2001)

will become established in the local host population, for instance by affecting the survivorship of the parasite's infective stages. We might thus expect parasite species richness to be higher in parasite-friendly habitats. Several studies have examined the association between abiotic features of lake habitats and the richness of parasite component communities of fish hosts; many patterns emerge from these studies, but whether they are merely spurious correlations or true cause-and-effect relationships is not always clear (Hartvigsen and Halvorsen 1994). Several studies have reported correlations between physical features of lakes, such as surface area or depth, and helminth parasite species richness of fish hosts across several host populations (e.g., Kennedy 1978; Marcogliese and Cone 1991b; Bergeron et al. 1997). In these cases, it may not be the lake features *per se* that influence parasite species richness, but their indirect effects on the availability of other host species for the completion of the parasite life cycles. In other cases, however, direct causal mechanisms can be identified. For instance, low pH values are often associated with low parasite species richness in freshwater fish populations, mainly because of the direct effect of acidity on the availability of calcium for snails: where there are no free calcium ions available, there are few or no snails, and thus all trematode species requiring snail intermediate hosts go locally extinct (Curtis and Rau 1980; Marcogliese and Cone 1996; Halmetoja et al. 2000). Similarly, water conductivity correlated negatively with the richness of the epibiont fauna on *Daphnia* across rock pools, presumably because of the direct effects of salinity on these external symbionts (Ebert et al. 2001).

In terrestrial habitats, environmental factors can also cause variation in parasite species richness among populations of the same host species. For example, *Anolis* lizards in the Caribbean islands harbor depauperate parasite assemblages in dry habitats and richer ones in habitats with higher levels of humidity (Dobson and Pacala 1992; Dobson et al. 1992). Only parasite species most tolerant of dry conditions occur in dry habitats, whereas humid habitats allow other species to persist as well. In the same vein, Krasnov et al. (1997) found that habitat types that allowed a favorable microclimate for fleas in their hosts' burrows were associated with higher species richness of fleas on desert rodents than other, less favorable habitats.

One of the greatest dichotomy of habitat types on Earth is the aquatic versus terrestrial divide. A glance at surveys of helminth parasites from aquatic and terrestrial hosts reveals a clear pattern: in any large group of vertebrates, aquatic species are hosts to more species of helminths than their terrestrial counterparts (Figure 4.10). There are a few exceptions, such as pelagic seabirds that usually harbor depauperate parasite assemblages (Hoberg 1996), but generally speaking, aquatic vertebrate hosts are exploited by a richer assemblage of parasites than their terrestrial relatives (Bush et al. 1990). Is there a causal link between living in an aquatic environment and accumulating many species of parasites over evolutionary time? The only way to answer this question is to restrict all comparisons between aquatic and terrestrial hosts to pairs of lineages derived from a common ancestor that have diverged in terms of habitat. If aquatic host taxa have consistently more parasites than their terrestrial sister taxa, once corrections have been made for confounding variables, then we could conclude that living in water *per se* has an effect on the richness of parasite faunas. There are actually few cases of habitat divergence in the evolutionary history of the main vertebrate lineages, and this places limits on the statistical power of proper comparative tests. Not surprisingly, both tests of the role of habitat in the evolution of parasite faunas of vertebrates have yielded no significant results (Poulin 1995c; Gregory et al. 1996). All we can say on this issue at the moment is that aquatic vertebrates harbor more parasites than their terrestrial counterparts, but that this may have nothing to do with their aquatic lifestyle (see Box 14.1 in Bush et al. 2001).

Host Diet

For parasites acquired by ingestion, such as most helminths in vertebrate hosts, what the host eats will determine the kinds of parasite infective stages that may colonize the host. Therefore both the breadth and composition of the host diet should influence parasite species richness. This simple prediction has proven difficult to test because it is not easy to classify host diet in a quantitative, or even qualitative, way that allows comparisons among different host species. Two host species may have the same diet breadth (e.g., they both consume 12 different types of prey items), but the identity of these items may be different. Alternatively, they may consume the same prey items, but in different relative proportions. How can this information be combined into an index that allows interspecific comparisons? Despite these difficulties, the role of host diet has been assessed in several studies, each using a crude index. These indices generally quantify the relative contribution of animal food to the overall diet, since eating animal prey exposes the

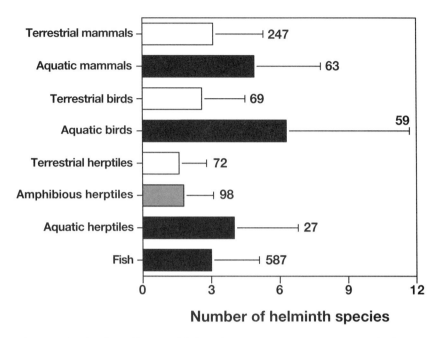

Figure 4.10. Number of intestinal helminth parasite species (mean ± standard deviation) per host population in different groups of vertebrates. Numbers next to the bars indicate the number of studies on which the estimates are based. The data include only helminth species with a prevalence of at least 10% in the host population, and thus exclude rare parasite species. Still, helminth species richness is generally higher in aquatic vertebrates (black bars) than in their terrestrial counterparts (open bars). Data include estimates from 245 species of fish hosts, 112 species of herptiles (amphibians and reptiles), 84 species of birds, and 141 species of mammals. (Data from Bush et al. 1990)

host to more helminths transmitted via intermediate hosts (see Bush 1990 for discussion). Some studies have found that host diet is an important determinant of parasite species richness (e.g., Bell and Burt 1991; Guégan and Kennedy 1993; Morand et al. 2000), whereas other studies did not (e.g., Gregory et al. 1991; Poulin 1995c; Watve and Sukumar 1995). Therefore host diet may be important in some cases, but not in all cases. Clearly, it is not an overriding factor in many systems, such as in the case of the gastrointestinal helminths of zebras and horses (Krecek et al. 1987; Bucknell et al. 1996), where a high parasite species richness is achieved by hosts with a rather narrow plant diet.

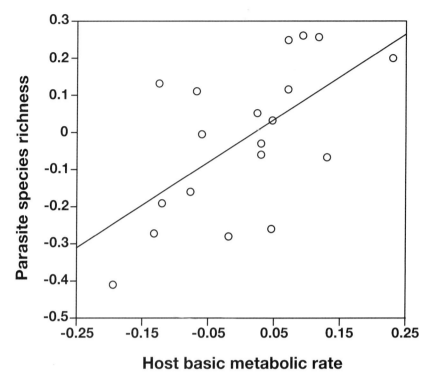

Figure 4.11. Relationship between the species richness of endoparasitic helminths and host basic metabolic rate across species of mammalian hosts. Points are phylogenetically independent contrasts derived from log-transformed data on 23 host species, corrected for confounding variables such as sampling effort and host body mass. (Data from Morand and Harvey 2000)

Host Metabolism

Regardless of what the host eats, the rate at which it eats and the amount of food material ingested should relate positively with the rate at which infective stages of parasites are acquired. Since the rate at which the host ingests parasite larvae is a determinant of the parasite's basic reproductive number, R_0 (ingestion rate is the parameter β that defines H in the equation for R_0; see Chapter 3), we might expect host species with a high ingestion rate to harbor more parasite species. Hosts that ingest parasite larvae at a high rate will allow more parasite species to reach the threshold R_0

value (i.e., $R_0 \geq 1$) necessary for their persistence. In addition, greater rates of food consumption by the host may ensure a greater supply of nutrients and energy for its parasites, perhaps allowing more of them to coexist. Thus we can predict that host species with a high metabolic rate, which have a high rate of food consumption, will harbor more species of intestinal helminths than host species with lower metabolism (see Gregory et al. 1996).

At a broad level, differences in parasite species richness between endothermic and ectothermic vertebrates are well documented, and suggest that endotherms have richer parasite faunas (Kennedy et al. 1986). Of course, on such a broad level, other factors may also play roles. For instance, the greater morphological differentiation of the intestine of endotherms might provide a greater variety of niches available for parasite occupation than would be found in fishes or herptiles (Kennedy et al. 1986). The role of metabolism *per se* must therefore be evaluated within a given group of vertebrates. The only test of this kind was performed by Morand and Harvey (2000), who found that after controlling for confounding variables, the basic metabolic rate of mammalian hosts correlated positively with the species richness of their parasitic helminths (Figure 4.11). This result provides support for a potential role of host metabolism in determining parasite species richness. As with other correlational evidence, the causal arrow can be turned around: it may be that high metabolic rates evolve in host species exploited by many parasites because of the high cost of the immune responses they mount against these invaders. Unraveling whether high metabolic rate is the cause or consequence of high parasite species richness will not be straightforward, especially at the interspecific level.

Host Genetics

Mechanisms of host defense such as the immune system have a genetic basis, and thus host genetics can determine how susceptible a host individual, population, or species is to invasion by parasite species. Evolutionary arguments have been proposed to suggest that parasites exert frequency-dependent selection on their hosts, favoring rare host alleles that may confer greater resistance against widespread parasites (Hamilton 1982). Genetic resistance against parasites can quickly become obsolete because parasites evolve at higher effective rates than their hosts, tracking the most common host genotypes: hosts with rare genotypes may be at a temporary advantage.

Thus the presence of parasites should maintain greater genetic variability among host individuals than would occur in their absence. Indeed, there is good field evidence that genetic variability is a major determinant of the risk of infection by particular parasites among host individuals (Dybdahl and Lively 1998; Liersch and Schmid-Hempel 1998; Paterson et al. 1998; Coltman et al. 1999).

At the level of the host population or host species, however, different patterns might be predicted. On the one hand, parasite species richness and host genetic variability can be positively correlated across host species if parasites are strong selective agents that have driven high genetic variation in their hosts, as argued above. This pattern has been found for monogenean ectoparasites of African freshwater fishes; host fish species with higher levels of genetic heterozygosity had higher parasite species richness than fish species with lower genetic variability (Pariselle 1996). On the other hand, if indeed genetic variability is advantageous for hosts in their fight against infection, parasites may be more successful at colonizing host populations or host species that show low levels of genetic variation, for whatever reason (see Meagher 1999). This would result in a negative relationship between host genetic variability and parasite species richness, with host species that are genetically homogeneous accumulating parasite species at a higher rate than related but genetically more variable host species. In accordance with this expectation, Poulin et al. (2000) reported a negative correlation between helminth species richness and host heterozygosity among North American freshwater fish species. Similarly, across species of ants, parasite species richness correlated negatively with host genetic variability; that is it correlated positively with average within-colony relatedness (Schmid-Hempel 1998; Schmid-Hempel and Crozier 1999). Clearly, host genetic variation is important, but whether it is a cause or consequence of parasite species richness remains to be settled.

Other aspects of the host genome may also be important. Guégan and Morand (1996) have found that, among African freshwater cyprinid fish, polyploid species harbor a richer fauna of ectoparasitic monogenean parasites than their diploid relatives. The interpretation of this result is as difficult as those above. Either polyploidy confers a selective advantage against parasites and is selected in fish species exploited by many parasites, or polyploidy disrupts host defenses and allows more parasite species to colonize the host.

Is Parasite Biodiversity Neutral?

The search for ecological or host-related correlates of parasite species richness is founded on the assumption that species richness is not randomly distributed among host species or geographic areas, but that it follows predictable patterns. Recently, some ecologists have championed an alternative view, the so-called neutral theory of biodiversity (Bell 2001; Hubbell 2001). This theory seeks to predict both the number of species found in a community and the distribution of the relative abundances of those species. It rests on the assumption that every individual in every species in a biological community is roughly identical, and that the total abundance of all species is fixed. All variation in species richness or their relative abundances occur because of purely random variation in births, deaths, migration, and speciation. The theory is remarkably accurate in its predictions about the properties of many ecological communities, particularly regarding the abundances and species richness of trees in tropical forests (Hubbell 2001). Individuals of different tree species often appear competitively equivalent, and the theory works superbly in situations where all individuals are equal (see Hubbell 2001; McGill 2003; Nee and Stone 2003; Volkov et al. 2003).

Could the neutral theory apply to parasite assemblages? The obvious answer must be no (see Poulin 2004b). There is no general evidence of competitive equivalence between parasite species. In assemblages of gastrointestinal helminth parasites in vertebrate hosts, for instance, cestodes often have body sizes several orders of magnitude larger than trematodes or nematodes that coexist with them; it would often be impossible for cestodes to occur at the same abundances as trematodes, simply because of space restrictions. In addition, there is much evidence of strong competitive interactions among enteric helminths. The presence of one species typically causes a functional shift in resource use by a second species (e.g., Holmes 1961; Bush and Holmes 1986b; Patrick 1991), or a numerical decrease in the numbers of individuals of the second species or in their fecundity (e.g., Silver et al. 1980; Dash 1981; Holland 1984). Frequently, these competitive interactions are asymmetrical, with one species clearly superior in competition and the other bearing the brunt (Poulin 1998a). It is impossible to speak of all species being equal in such assemblages of parasite species (Poulin 2004b), and thus we must look for other explanations for variation in parasite diversity.

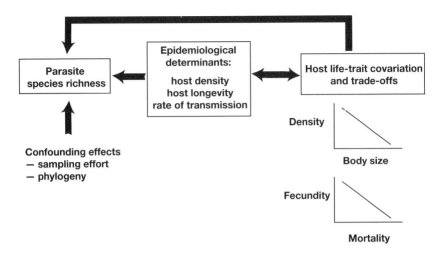

Figure 4.12. Schematic illustration of the interrelationships between parasite species richness, epidemiological parameters, host life history traits, and confounding variables.

Conclusions

Most of the empirical evidence reviewed in this chapter concerns metazoan parasites (helminths and arthropods) of vertebrate hosts. Little is known about patterns in species diversity of other groups of parasites on other groups of hosts (for an example, see Poulin and Mouritsen 2003). If we expand our definition of parasites to include other organisms that feed on hosts without, or prior to, killing them, we could include classical studies of diversity patterns in phytophagous insects on plant hosts (Lawton 1983; Strong et al. 1984) and in parasitoids of insect hosts (Hawkins 1994). Despite the fact that insect parasites are quite different from metazoan parasites of vertebrates, large-scale macroecological studies of these insects have revealed patterns close to those described here, suggesting the actions of similar processes of diversification.

Although certain host traits play a significant role in a range of systems, it is clear that there is no universally important ecological determinant of species richness in either parasite component communities or parasite faunas. Specific features of the transmission mode of parasites or of host biology can dictate which of the ecological variables reviewed here will play the biggest role in the diversification of the parasite faunas of a group of hosts. In the end, no single factor has acted alone over evolutionary time: parasite species richness is the outcome of interactions between a constellation of processes affecting colonization, speciation, and extinction rates of parasite lineages. We propose a complex framework for studying the determinants of parasite species richness, outlined in Figure 4.12. After accounting for the confounding influence of sampling effort and host phylogeny, epidemiological parameters are likely to be the best predictors of parasite species richness. Some of these determinants, however, such as host population size (or host population density) and host longevity, are related to each other via allometric covariation and trade-offs among host life-history traits; the latter can also influence parasite species richness directly (Figure 4.12). Only multivariate approaches will prove informative.

Without quantitative predictions and comparisons with appropriate null patterns generated by neutral models of the random assembly of parasite faunas (see Bell 2001; Hubbell 2001), the ecological processes underlying parasite species richness remain speculative. Clear, quantitative predictions would also make it easier to determine the direction of causality for many observed relationships involving parasite species richness. We have achieved a deeper understanding of why certain animal species are hosts to more parasite species than other animal species, but because these kinds of questions cannot be tackled experimentally, we may never achieve the final answer.

5

Parasite Features and
Parasite Diversification

Species richness is almost always distributed unevenly among clades within any given higher taxon. Typically, one or a few clades are numerically dominant—they include a disproportionate number of species relative to other related clades (Dial and Marzluff 1989). For instance, passerines include many more species than all other orders of birds (almost 30) put together, and rodents include well over a third of all mammalian species, although they represent only one of 19 orders of mammals. Similar examples could be drawn from lower taxonomic levels or from any other phylum, including parasite taxa. For example, among monogeneans, two families, Dactylogyridae and Gyrodactylidae, account for a disproportionate amount of the group's biodiversity (Cribb et al. 2002b). Within families in any taxa of parasites, a similar pattern often emerges (Figure 5.1). This common trend cannot be ascribed entirely to taxonomical artifacts, that is, to widespread asymmetries in the tendency of systematists to either split or lump taxa. Instead, it is to some extent an indication that the few dominant clades experience more speciation events and fewer extinctions than other clades at the same taxonomic level. Simulation studies have shown that extreme overdominance of one or a few taxa is not expected from null models of species diversification among higher taxa in which all contemporaneous species are assigned the same probability of speciation (Dial and Marzluff 1989; Slowinski and Guyer 1989; Nee et al. 1996). Clearly, there must be real differences in evolutionary success among related clades. These differences are most likely linked to intrinsic properties of the species within clades, with some clades possessing traits that can either promote speciation or reduce

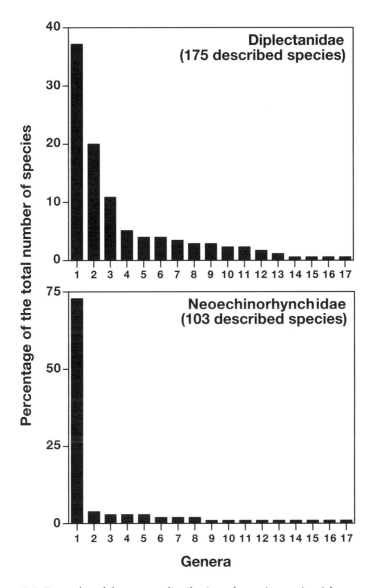

Figure 5.1. Examples of the uneven distribution of parasite species richness across related parasite taxa. The figure shows the relative number of described species in each genus of two families, the Diplectanidae (Trematoda) and the Neoechinorhynchidae (Acanthocephala). Each family contains 17 genera, and they are ranked in decreasing order of species richness. Clearly, the most speciose genera account for most of the species richness in each family. (Data from Amin 1985; Desdevises et al. 2001)

the risk of extinction, and related but less-speciose clades not possessing these traits (Marzluff and Dial 1991; Slowinski and Guyer 1993; Barraclough et al. 1998).

Parasitism *per se* may be one of those features favoring diversification. For instance, there are clearly many more parasitic flatworms than free-living ones within the phylum Platyhelminthes (Barnes 1998), and families and genera of parasitic nematodes are relatively more diversified than those of their free-living counterparts within the phylum Nematoda (Morand 1996a). Similarly, taxa of phytophagous insects, which fit in a broad definition of parasites, tend to be more speciose than their sister taxa that are not parasitic on plants (Mitter et al. 1988). No rigorous attempts have been made to compare the species richness of taxa with parasitic modes of life with the richness of their free-living sister taxa, and it is thus too early to say whether being a parasite facilitates diversification. In contrast, more is known about the potential causes for the greater diversity of certain parasite taxa relative to others.

This chapter looks at the diversification of parasite lineages and how features of parasites within clades can influence rates of speciation or extinction. Whereas in the preceding chapters we focused on the distribution of parasite species richness among host taxa, in this chapter we examine the distribution of parasite species richness among parasite taxa. We begin with a discussion of sympatric, or intrahost, speciation in parasites, and then proceed to an investigation of how parasite body size and other parasite features are linked with their diversification.

Intrahost Speciation and Congeneric Parasites

Intrahost speciation is the process by which a parasite lineage speciates within a host lineage, without an accompanying host speciation event (see Chapter 4). In a situation with strict cospeciation of parasites each time their host speciates, and no host switching, beginning with one ancestral host species harboring one parasite species, we eventually get n host species harboring n parasite species. If intrahost parasite speciation can also take place, we will obtain n host species harboring $>n$ parasite species. If host switching were infrequent, this would be the only way in which new parasite species could be generated without a concomitant increase in host diversity.

According to some reconstructions of host–parasite coevolutionary history, intrahost speciation events can be relatively common (e.g., Paterson et al. 1993; Paterson and Gray 1997). These reconstructions, however, are based on several assumptions built into the algorithm used to assess the congruence between host and parasite phylogenies. We must therefore look for other lines of evidence supporting the existence and commonness of intrahost speciation. As a speciation process, intrahost speciation is not much different from classical sympatric speciation (i.e., speciation in the same geographic area without physical isolation and barriers to gene flow). As an alternative to allopatric speciation, sympatric speciation has long been considered an unlikely or secondary source of diversity. Recent developments, however, have shown that it may be a common evolutionary event under a wide range of conditions (Bush 1994; Via 2001). Both recent theoretical models (Dieckmann and Doebeli 1999; Kondrashov and Kondrashov 1999) and empirical evidence (Via 2001; Berlocher and Feder 2002) have shed a more favorable light on sympatric speciation. Generally speaking, a divergence in resource use can be strengthened by ecological selection (as opposed to sexual selection) against hybrids or morphological intermediates (Schluter 2000). In particular, a range of studies on phytophagous insects have provided support for a sympatric speciation mode in which new insect species are formed locally after switching host plants (e.g., Bush 1969, 1994; Tauber and Tauber 1989; Berlocher and Feder 2002). Although sympatric on a geographic scale, this sort of speciation is no different from speciation after a host switch (see Chapter 4), and is not really the same as intrahost speciation. The latter involves sympatric speciation on a much smaller spatial scale: speciation on the same host species but after a switch to a different niche within the host.

Defining sympatry and the scale at which one considers species to be sympatric is particularly important in studies of parasite diversification (McCoy 2003). Among parasites, there are examples of sympatric speciation on the larger scale considered in studies of phytophagous insects. For instance, inherited mechanisms of host preferences serve to maintain reproductive isolation in the field between two sympatric species of copepods parasitic on flatfish, although they can mate and produce fertile hybrids in the laboratory (de Meeûs et al. 1995). Similarly, heritable asynchrony in cercarial emergence from snail intermediate hosts has been driving the sympatric speciation of a schistosome trematode using two different definitive hosts (Théron and Combes 1995). These are cases of parasite speciation on different but geographically sympatric host species. This process is believed to be common in the evolution and diversification of parasites, to the extent that it has serious implications for our understanding of what exactly represents a parasite species (de Meeûs et al. 1998; Kunz 2002).

What about true intrahost speciation in parasites? Although their evolutionary origins are difficult to ascertain precisely, congeneric species of parasites exploiting the same host species are widely thought to represent the outcome of intrahost speciation events. The co-occurrence of two closely related parasite species on the same host has a limited number of possible explanations: either an ancestral parasite species went through intrahost speciation, or the two related parasite species came together after one or two host-switching events among related hosts (Figure 5.2). Thus, the outcome of intrahost speciation may be indistinguishable from that of secondary contact between allopatrically derived species after host switching; it is often difficult to discount "the ghost of allopatry past" (Via 2001). It is also possible that intrahost speciation occurs in temporarily allopatric host populations, which subsequently are reunited and swap their new parasites. Without knowledge of the relative likelihood of these different evolutionary events, it is difficult to say whether conspecific parasite species sharing the same host are indeed the product of intrahost speciation.

In spite of these uncertainties, many well-known systems in which multiple congeners occur on the same hosts are considered—based on several lines of evidence—to be monophyletic assemblages issued from high rates of intrahost speciation and radiation (Schad 1963; Inglis 1971; Kennedy and Bush 1992). These assemblages of congeners are seen as the parasite equivalents of the species "flocks" seen in certain free-living animal communities (see Mayr 1984). They include, among others, oxyurid nematodes in tortoises (Schad 1963), strongyloid nematodes in elephants and horses (Inglis 1971; Bucknell et al. 1996), cloacinid nematodes in kangaroos (Beveridge and Spratt 1996; Beveridge et al. 2002), and dactylogyrid monogeneans on freshwater fishes (Kennedy and Bush 1992; Simkova et al. 2004). In some of these cases, the main driving force behind the apparent intrahost multiplication of parasite species seems to be the partitioning of resources, as indicated by the great differentiation of feeding structures (Beveridge and Spratt 1996). These systems may be the most dramatic examples of intrahost speciation and thus the most easily recognizable, although the phenomenon could be widespread in other host–parasite associations (e.g., Bray 1986).

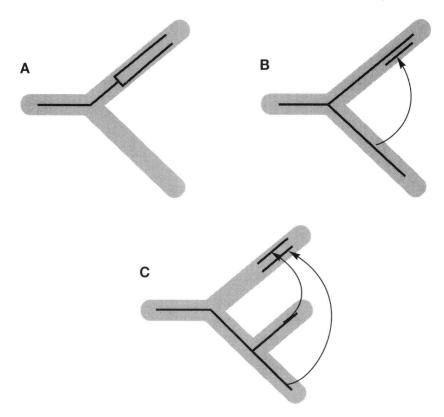

Figure 5.2. Three scenarios leading to the co-occurrence of parasite sister species in the same host species: (A) intrahost speciation, (B) one host switch to a closely related host species, and (C) two independent switches to another related host species. The parasite phylogeny (black line) is superimposed on the host phylogeny (shaded area); host switches are indicated by arrows. Switches to more distantly related host species are also possible, but not illustrated here.

Two points deserve further discussion. First, are co-occurring congeneric parasite species always the outcome of intrahost speciation? The answer is no, of course. Congeneric bird species on the same island are not necessarily more closely related to each other than to other congeneric species on the mainland (Coyne and Price 2000), and the same is true of parasites on the same host. For instance, among polystomatid monogeneans parasitic in freshwater turtles, congeneric species infecting the same site in different host species, even if the latter are geographically or phylogenetically distant, are more closely related to one another than to congeners infecting different sites on the same host species (Littlewood et al. 1997). Similarly, in the monogenean genus *Gyrodactylus* parasitic on freshwater fishes, frequent host switching across large phylogenetic distances has played at least as big a role as intrahost speciation in generating many co-occurring congeneric species (Zietara and Lumme 2002). Thus we cannot simply take the presence of multiple congeners on the same host as evidence of past intrahost speciation.

Second, assuming that most congeneric parasite species are indeed the product of intrahost speciation, are rates of speciation determined by host features or by parasite features? Surely certain hosts provide their parasites with more opportunities for intrahost speciation than other hosts, for instance because of the broader range of niches they offer for parasites. There are indeed patterns in the occurrence of multiple congeners among host taxa. In assemblages of gastrointestinal helminths of vertebrates, parasite genera represented by a single species are almost always more frequent than genera represented by two or more species (Kennedy and Bush 1992). Multiple congeners are not rare, however, and they are generally more common in endothermic vertebrate hosts (birds and mammals) than in ectothermic vertebrates (Figure 5.3). Other patterns also emerge from an examination of the available data. For example, aquatic bird species tend to harbor more congeneric helminth species than terrestrial birds, and large-bodied mammal species harbor more congeneric helminths than small-bodied ones (Poulin 1999a). None of these patterns survives after appropriate corrections for potential phylogenetic influences, however (Poulin 1999a). This suggests that the observed patterns reflect the current distribution of congeneric parasite species without shedding light on the evolutionary processes responsible for the within-genus diversification of parasites. In other words, large mammals currently harbor more congeneric helminths than small mammals, but host body size *per se* is not the driving force behind this pattern. An analysis of the association between host body size and the occurrence of congeneric species among monogenean ectoparasites of fish has also found no relationship between host size and the relative number of congeners per fish species (Poulin 2002).

On a larger scale, there may be differences between higher host taxa in how likely their parasites are to undergo intrahost speciation. These differences can become apparent when looking at how the number of higher taxa

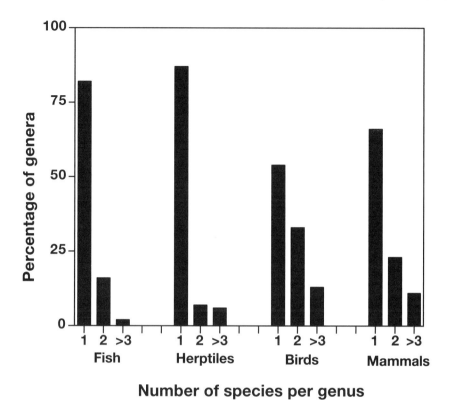

Figure 5.3. Frequency distribution of genera of intestinal helminth parasites represented by one or more species within a host population, for different groups of vertebrate hosts. Data are from 245 fish host species, 112 herptile species (amphibians and reptiles), 84 bird species, and 141 mammal species. (Data from Kennedy and Bush 1992)

(genera or families) per parasite assemblage relates to the number of species in these assemblages. Recently, Enquist et al. (2002) have shown that the number of genera or families in woody plant communities is a simple power function of species richness; the exponents of the power functions they derived are remarkably consistent over a range of scales. In parasite assemblages of host populations or species, increases in species richness occur in only two ways, either from within the assemblage (intrahost parasite speciation) or from outside (via host switching). If the number of genera in parasite assemblages is a simple power function of species richness, we can gain some insight into which of these two processes has been most important in the diversification of parasite communities. If we find an exponent of 1 for the relationship between the number of parasite genera and the number of parasite species across several comparable parasite assemblages, then host switching has been the main cause of diversification, because each species present in a community is taxonomically independent of the other species, and must therefore have had a separate origin. If, on the other hand, the exponent is much less than 1, then several species belong to the same genus or genera, suggesting that they have a common ancestor and that they have radiated (by intrahost speciation) after a host switch by that ancestor. The relationship between the number of genera and the number of species in communities of helminth (cestodes, trematodes, acanthocephalans, and nematodes) endoparasites in fishes or birds is almost one for one, whereas among mammal hosts the number of parasite genera increases significantly more slowly as a function of species richness (Figure 5.4). Given that helminth communities in all vertebrates consist of the same four large taxa, it is unlikely that the weaker relationship in mammals is due to artifacts of taxonomic classification. These findings suggest that in mammals in general, rates of within-host parasite speciation may be higher than in other vertebrate hosts. The reasons for this apparent difference remain unclear. Attempts to link host features with intrahost parasite diversification are still few. Apart from the pattern in Figure 5.4, robust trends have yet to emerge, a fact suggesting that parasite diversification may be driven to a large extent by properties of the parasites themselves.

Parasite Body Size and Parasite Diversification

The last section focused on intrahost parasite speciation and showed that, although this evolutionary process is plausible, it is not the only—or even the main—source of parasite species diversity. New parasite species are also produced via cospeciation with hosts and speciation after a host switch. Because certain parasite taxa are much more speciose than others, we might speculate that species-rich taxa possess some key features that facilitate speciation and thus diversification by any of these speciation processes. So what are these features?

Looking back to epidemiological models (Chapter 3), it is clear that they can be used to predict the rates of speciation and diversification in different parasite taxa. If parasite species with a high basic reproductive num-

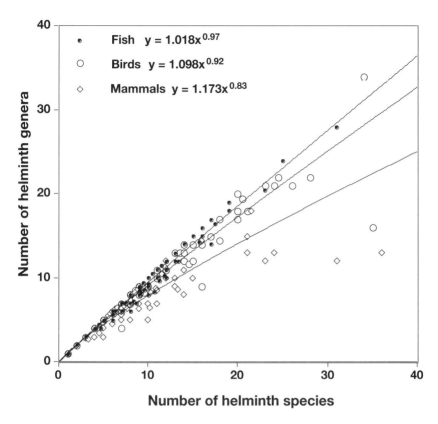

Figure 5.4. Relationship between the number of parasite species and the number of parasite genera per host species for 76 bird species, 107 fish species, and 114 mammal species. The relationship follows a simple power function that is lower for mammals than for fishes or birds. Where data were available for more than one population of host species, values were averaged to obtain species values. (From Mouillot and Poulin 2004)

ber, R_0, can easily invade, spread, and persist in a host population over ecological time, they may also be capable of colonizing and persisting in new host species over evolutionary time. With the subsequent cospeciation of hosts and parasites, the result is a greater likelihood of diversification in parasite taxa with high R_0. One testable prediction would be that parasite taxa, say genera, characterized by high R_0 values should be more speciose than related taxa characterized by lower R_0 values. We have insufficient data on R_0 from a range of parasite taxa to test this now, but integrating epidemiological information with studies of parasite diversity is clearly a fruitful avenue for future research.

The only life-history trait for which data are available for numerous parasite species is body size. In most organisms, body size covaries with most other key life-history traits, such as fecundity, metabolic rate, age at maturity, and lifespan (Peters 1983; Schmidt Nielsen 1984). The same is true for parasites (e.g., Skorping et al. 1991; Morand 1996b; Poulin 1996c, 1998a; Trouvé et al. 1998). Many parameters included in the equation for R_0 are therefore related to parasite body size, and this trait is thus an ideal candidate in our search for features of parasites that may affect their rates of speciation and diversification.

In most higher groups of organisms, body size distributions are right-skewed, even on a logarithmic scale, such that the smaller size classes, but usually not the smallest ones, include more species than the larger size classes (Van Valen 1973; Dial and Marzluff 1988; Fenchel 1993; Blackburn and Gaston 1994a; Brown 1995). A range of mechanisms are probably acting synergistically to promote the proliferation of small-bodied organisms (reviewed in Brown 1995; Gaston and Blackburn 2000; Kozlowski and Gawelczyk 2002); only those most relevant to parasites are mentioned here. Ultimately, the numbers of species in different size classes depend on size-related rates of speciation and extinction; size classes with the highest net rate of diversification (speciation rate minus extinction rate) will have the highest species richness (Dial and Marzluff 1988; Maurer et al. 1992). So why should small organisms achieve higher net rates of diversification than large-bodied ones? Two of the many proposed answers stand out from the rest because of their generality and the level of empirical support they received (see Gaston and Blackburn 2000). First, the fractal nature of most habitats means that small organisms will perceive more available space and niches in a given area than a large organism would (Morse et al. 1985; Shorrocks et al. 1991; Gunnarsson 1992). It may simply be a matter of scale: perceived habitat heterogeneity is size-dependent, and more small species can coexist than large ones in a given area because small species can subdivide the environment more finely. Second, rates of energy acquisition and conversion into offspring are also size-dependent, peaking at relatively small sizes (Brown et al. 1993; but see Kozlowski 2002). Organisms close to the optimal body size for energy acquisition should therefore be at an advantage, facing a lower risk of extinction when resource availability declines

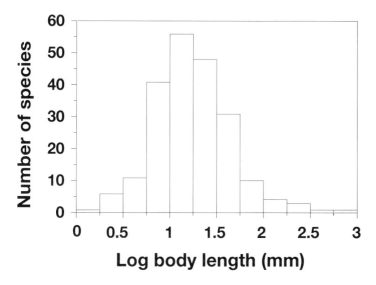

Figure 5.5. Frequency distribution of body lengths of 810 nematode species parasitic on mammals. (Data from Morand and Poulin 2002)

and having greater success at colonizing new areas where they may speciate. These and other mechanisms, combined with the shorter generation times of small-bodied organisms (Marzluff and Dial 1991), should lead to the proliferation of small taxa relative to larger ones.

The body size distributions of parasites also tend to be right-skewed, but not as consistently as those of free-living animals (Poulin and Morand 1997). Typically, the body size distributions of species within higher parasite taxa show a slight to moderate skew, as in nematodes parasitic on mammals (Figure 5.5). Only the monogeneans show a strikingly different pattern: their body size distribution is right-skewed, but bimodal (Figure 5.6). There may be real biological explanations for this pattern (Poulin 1996d, 2002), but the most likely reason is that the Monogenea are apparently not monophyletic, and instead consist of two distinct groups, Monopisthocotylea and Polyopisthocotylea, each with a unimodal, right-skewed body size distribution (Poulin 2002). An overview of skewness in parasite body size distributions reveals some intriguing patterns. For instance, interspecific body size distributions of male parasites differ from those of their female counterparts in most dioecious parasite clades (Poulin and Morand 1997). These differences probably reflect sexual differences in optimal body

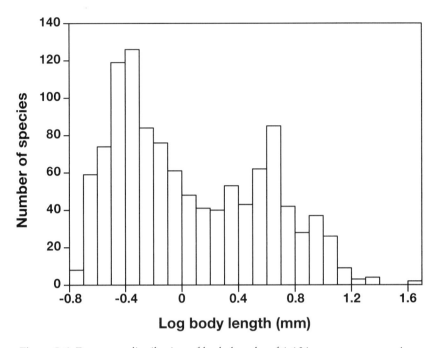

Figure 5.6. Frequency distribution of body lengths of 1,131 monogenean species parasitic on fishes. (Data from Poulin 2002)

size to maximize reproductive output. More importantly, body size distributions also appear to be influenced by the type of host exploited by parasites, and by the site of attachment on the host. Specifically, endoparasites generally tend to show more pronounced right-skewed distributions than ectoparasites (Poulin and Morand 1997). This result suggests that space constraints favoring small-bodied taxa may be more severe for internal parasites than for external parasites.

Ignoring these minor influences, it is clear that the distribution of parasite body sizes are generally no different from those of free-living organisms: within the size range exhibited by species belonging to a given parasite clade, there are more species toward the lower end of the size range than there are close to the upper bound. As discussed elsewhere (Chapter 1), recently discovered parasite species tend to be smaller than species described several decades ago. The size distributions we have now are only approximations: we should expect body size distributions of parasites to become increasingly right-skewed (Blackburn and Gaston 1994b). This strongly suggests that small-bodied taxa experience higher net rates of speciation than larger-bodied ones.

Using a simulation approach similar to that advocated by Brown et al. (1993), Morand and Poulin (2002) generated a distribution of nematode body sizes based on the assumption that nematode body sizes have been selected to optimize R_0. The host and parasite traits that determine R_0 (see Chapter 3) are all dependent on body size; using the known body mass distribution of mammal species, Morand and Poulin (2002) derived the expected frequency distribution of nematode body sizes that would optimize R_0 for these hosts. Remarkably, the simulated frequency distribution had the same mean and almost the same shape as the observed one (shown in Figure 5.5). Thus, selection acting on R_0 could have produced slightly right-skewed body size distribution in parasitic nematodes, and a similar process may be acting in other taxa.

The true acid test of the idea that small body sizes promote diversification would involve comparing species richness between sister taxa that differ in body size. If the smaller-bodied taxa consistently have higher species richness than their sister taxa with larger body sizes, then we have much stronger evidence in favor of size-dependent diversification. There are variants on the phylogenetically independent contrasts method (see Chapter 4) that are specifically designed to investigate the correlates of species richness, such as body size, while controlling for phylogenetic influences (Agapow and Isaac 2002; Isaac et al. 2003). A few proper comparative analyses of species richness as a function of body size have been performed on free-living organisms; the results indicate that body size is not a universal determinant of diversification (Gittleman and Purvis 1998; Gardezi and da Silva 1999; Orme et al. 2002a,b). The only investigation on parasites has focused on monogeneans (Poulin 2002), and revealed no associations between the number of described species per monogenean family and the mean body length of species within families (Figure 5.7). The same is true when using the mean number of species per genus as a measure of diversification rates within a family (Poulin 2002). The points for two small-bodied and diverse families, Dactylogyridae and Gyrodactylidae, fall in the upper-left corner of the plot, as we would expect if diversification rates were driven by body size (see Figure 5.7). However, the pattern disappears when all families are considered. Further such comparative tests are necessary to assess the overall role of body size in parasite diversification.

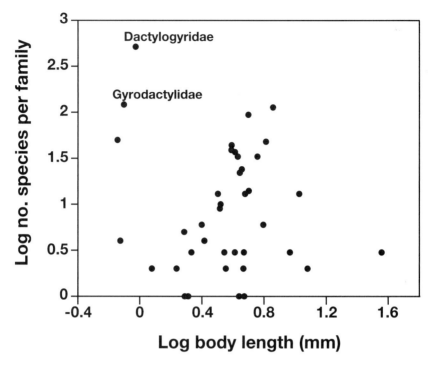

Figure 5.7. Relationship between the number of described species per monogenean family and the mean body length of species within families, across 39 monogenean families. The positions of two very speciose and small-bodied families, Dactylogyridae and Gyrodactylidae, are shown. (Data from Poulin 2002)

Other Parasite Features and Parasite Diversification

Because it correlates with all major life-history traits determining the fitness of parasites, body size is a prime candidate as a promoter of parasite diversification rates. Other properties of parasites may also affect their chances of speciating or going extinct. Among them, the number of hosts required to complete the life cycle and the specificity of parasites for given host species at each stage of the life cycle, appear particularly important. Here, we discuss these two parasite features and the role they may have played in generating the existing parasite diversity.

Perhaps the greatest constraint on parasite evolution, from the perspective of diversification, is the complexity of their life cycles. Many helminth parasite lineages have evolved complex cycles, requiring passage through a specific sequence of host species for successful development and maturation. The complexity of the life cycle may influence the number of opportunities for speciation and diversification. For instance, if a parasite population can use two or more species of definitive hosts after passage through an intermediate host, individuals within the population will have the possibility of specializing on one or the other definitive host species, after sharing the same intermediate host species. Such an opportunity for the formation of two distinct but sympatric parasite species would not exist if the parasite population had a simple cycle and used only a single host. A test of the effect of the life cycle on diversification in nematode parasites of vertebrates, however, did not support this prediction (Morand 1996a). The number of genera per family and the number of species per genus did not differ between nematode taxa with simple life cycles (i.e., a single host) and taxa with complex cycles (Figure 5.8). There is therefore no evidence that the complexity of the life cycle is a moderating factor in nematode diversification.

The situation may be different in platyhelminths. The phylum includes two diverse groups of endoparasites with complex life cycles (the trematodes and the cestodes), a large group of ectoparasites with a simple life cycle (the monogeneans), as well as a few small groups of free-living, symbiotic or parasitic worms. Brooks and McLennan (1993a, 1993b) have compared the species richness of the different groups of platyhelminths from a phylogenetic perspective. The high diversity of trematodes, cestodes, and monogeneans is independently derived—it evolved independently in each taxon. Based on several lines of evidence, Brooks and McLennan (1993a, 1993b) concluded that an adaptive radiation had occurred only in monogeneans and not in the other two groups. The monogenean diversification was attributed to a key evolutionary innovation: the loss of one host species from the life cycle and a reversal to a simple, one-host cycle (Brooks and McLennan 1993a). This result would support the suggestion made earlier that complex life cycles may constrain parasite diversification. The conclusions of Brooks and McLennan (1993a, 1993b), however, depend entirely on their phylogenetic hypothesis regarding the relationships among platyhelminth taxa and the evolution of life cycles. Other hypotheses exist and lead to completely different interpretations (Rohde 1996). In addition, as pointed out by Cribb et al. (2002b), monogeneans differ from other parasitic platy-

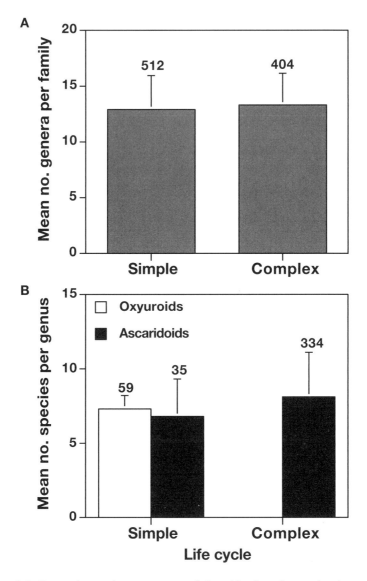

Figure 5.8. Comparisons of two measures of diversification, the number (mean ± standard error) of genera per family (A) and species per genus (B), of nematodes parasitic on vertebrates with either simple (one-host) or complex (≥two-host) life cycles. Data on species per genus are presented separately for oxyuroid nematodes, which all have simple life cycles, and ascaridoid nematodes, which have both types of life cycles. Numbers above bars are the numbers of genera or species used in calculating mean values. (Data from Morand 1996a)

helminths in many features other than the complexity of the life cycles, and determining which feature, if any, actually affects rates of diversification will be difficult.

Interestingly, within the trematodes, the digenean clade is much more diversified than its sister aspidogastrean clade, despite the latter usually having a simpler life cycle (two-host versus three-host for most digeneans). Similarly, within the digeneans, the family Didymozoidae is one of the largest, although most of its species have a four-host life cycle (Cribb et al. 2002b). More complex life cycles are therefore not an insurmountable obstacle to diversification. It is thus too early to assess the importance of the life cycle in promoting or limiting parasite speciation and diversification.

Turning now to host specificity, we can also make predictions regarding the relationship between the diversity of a taxon and the average host specificity of its member species. Parasites that are highly specialized (i.e., restricted to one or very few host species) should be more likely than generalist parasites to undergo speciation on a new host species once they colonize it. Because specialist parasites adapt quickly to changes in their immediate environment rather than maintaining a tolerance to a broad range of conditions, they should be more likely to produce a variety of new species each time they switch to new hosts (Brooks and McLennan 1993b). This scenario has also been proposed in the context of phytophagous insects that have an intimate association with their host plant similar to that between hosts and parasites (Futuyma and Moreno 1988). The only rigorous test of this prediction with parasites has examined the relationship between the mean number of host species exploited by parasites within a genus, and the number of species per genus, across genera of the monogenean family Diplectanidae (Desdevises et al. 2001). No significant association was found between host specificity and species richness across diplectanid genera, although many potentially confounding variables were taken into account. Since this is the only proper test to date of this hypothesis, it is too early to say any more about its validity.

Conclusions

Certain clades of parasitic animals are much more diverse than other, related clades. This pattern occurs within all higher taxa of parasites. Among those species-rich clades, the presence of several congeneric parasite species in the same host species suggests that even without switching hosts, parasites are probably capable of some forms of sympatric speciation. So what makes certain parasites more likely to proliferate into new species than others? We have reviewed studies that investigated promising features such as body size or aspects of the life cycle. Overall, the level of support for a role of any of these features in parasite diversification was rather low. The main reason for this is the lack of appropriate studies: there is simply not enough evidence to date to assess the many hypotheses proposed to explain rates of parasite diversification. Parasites are superb biological models for the study of ecological specialization, speciation mechanisms, and rates of diversification (de Meeûs et al. 1998). The main advantage of working with parasites is that the phylogeny of their hosts provides a precise historical record of the habitats used by parasites and their ancestors. We will need more evolutionary biologists to turn their attention to parasites to elucidate why some parasite lineages have been so successful.

6

The Biogeography of
Parasite Diversity

Biogeography is concerned with patterns in the distributions of species at spatial scales extending to the entire planet and the mechanisms that determine these distributions (Cox and Moore 1993; Brown and Lomolino 1998; MacDonald 2001). The number of overlapping species distributions in a given area determines local species richness. In previous chapters, much of the discussion focused on the partitioning of parasite diversity among host species, and not among geographic areas. In studies of species diversity of free-living organisms, however, the geographic component of diversity is usually of great interest (Ricklefs and Schluter 1993; Rosenzweig 1995). Applied to parasites, a biogeographic perspective raises interesting questions. For instance, are there regions of the world in which we find proportionally more parasite species? Has parasite diversification somehow been facilitated in these regions, and if so why? Although some trends in the biogeography of diversity have been demonstrated for parasitoid and gall-inducing insects (Hawkins 1994; Price et al. 1998; Mendonca 2001), parasites, in the strict sense used in this book, have not received much attention to date. Some studies have explored the relationship between local and regional richness in parasite assemblages; that is, how the parasite species richness in a given locality relates to the availability of parasite species within a larger area (e.g., Aho and Bush 1993; Kennedy and Guégan 1994; Barker et al. 1996). Because local parasite assemblages are established on much smaller spatial and time scales than the ones necessary for the evolutionary diversification discussed here, they reflect local ecological processes and not general biogeographic phenomena.

The main pitfall in biogeographic studies of relatively poorly known organisms such as parasites is the risk that a map of the distribution of their diversity merely reflects the variability in research activity in different parts of the world. Some correction for research effort is therefore necessary in using inventories of known species built from the literature. This is identical to the sampling effort problems discussed in Chapter 2. Even when a correction is made for uneven sampling, however, the results may be difficult to interpret (see Gibson and Bray 1994). This may be because many null hypotheses assume an even distribution of parasites and their hosts within geographic areas. Thus it is probably more informative to focus on a subset of parasites from a limited number of regions for which reliable information is available. The trends presented in this chapter are drawn from such studies, and for now they can be extended to all parasite taxa or all regions only by speculative extrapolation.

We begin by examining how the spatial distribution of parasite species richness is dependent on both the historical associations between parasites, hosts, and geographic regions and the current geographic distribution of their hosts. We then review the evidence for large-scale gradients in parasite species diversity and the mechanisms proposed to explain them. Finally, we consider the effect of the introduction of hosts to new geographic areas on the richness of their parasite assemblages.

Historical Biogeography and Parasite Diversity

Although this book focuses almost exclusively on quantitative aspects of parasite diversity rather than on qualitative ones (i.e., on the richness rather than the composition of parasite assemblages), we briefly cover historical events because they determine to a large extent not only which parasite species, but also how many parasite species are found in a particular area. The presence of certain species in an area has several possible explanations, involving both historical and ecological phenomena (Cox and Moore 1993; Brown and Lomolino 1998; MacDonald 2001). First, from the perspective of historical biogeography, current species distributions may be the result of past fragmentation of the biota of a larger region into isolated populations by new geographic barriers (vicariance biogeography), such as originally continuous land masses divided by oceanic straits through plate tectonics. On a smaller time scale, glaciation and sea-level fluctuations may also account for species distributions. Most organisms appear not to change their tolerances, but instead track suitable environments, with, for example, cold-intolerant species changing latitude as a function of glaciation. As an alternative to vicariance biogeography, species may have dispersed from various centers of origin by their own means and across pre-existing barriers. In the end, current patterns of species distributions are likely to be the product of both vicariance and dispersal. A second explanation involves ecological biogeography, which attempts to understand species distributions from the perspective of short-term, small-scale ecological processes (e.g., Hengeveld 1990). This theory has ties with the studies of parasite diversity covered in Chapter 4. Overall, biogeography aims to understand the patterns and processes in the distributions of species and thus to understand the geographic distribution of species richness.

The relationship between historical biogeography and host–parasite co-evolution has been investigated extensively in a handful of systems (see Brooks 1992; Hoberg 1992, 1995, 1996, 1997; Klassen 1992; Brooks and McLennan 1993b; Carney and Dick 2000; Choudhury and Dick 2001; Hoberg and Klassen 2002). This profusion of studies indicates that historical biogeographic events are important for current parasite diversity. These studies used phylogenetic analyses to show that vicariance (in which ancestral host and parasite populations are fragmented by emerging barriers) has regularly led to allopatric speciation in the parasites as well as in their hosts. The local species richness of the parasite assemblage, at the level of the component community, is not directly affected by this speciation, but the global species richness of parasite taxa increases each time a new parasite species arises. The effectiveness of barriers to parasite dispersal, and therefore the importance of vicariance, is illustrated by studies of external parasites of marine fishes. These parasites do not require intermediate hosts and are simply transmitted by free-swimming stages; they are rarely capable of crossing large oceanic distances to colonize related hosts in distant areas, leading to the allopatric evolution of different assemblages of parasite species (Hayward 1997; Rohde and Hayward 2000).

Although historical biogeographic events can promote some forms of parasite speciation and increase global parasite diversity by splitting an ancestral species into several new ones, they are not normally important determinants of local parasite species richness (richness in a given host population). There are exceptions, off course. For example, Hoberg (1992) proposed that historical biogeographic events linked with glaciation periods may have caused the diversification of cestodes in seabirds and pinnipeds. Periodic range contraction during glaciation isolated host species in refugial centers, followed by their expansion into postglacial habitats. Over the past few million years, these dramatic climatic fluctuations caused the regular break up of host populations into small refugia and thus promoted relatively rapid and extensive diversification of their cestode faunas (Hoberg 1992). Generally, however, if parasite species richness changes in an area over time, the cause must lie in properties of the area itself. This aspect of the biogeography of parasite diversity is of greater interest here, and we now turn to the factors associated with hotspots of parasite diversity.

Parasites Are Where the Hosts Are

On a moderate geographic scale (say, hundreds to thousands of kilometers), variability in habitat quality will create parallel variability in host species richness. Within species-rich patches of hosts, we should find more parasite species than outside the patches. All else being equal, then the more host species exist in one locality, the more parasite species should exist there as well. If one lake is inhabited by twice as many fish species as a second lake, we would expect it, on average, to also contain twice as many parasite species. Because parasites are intimately associated with their hosts, the spatial distribution of host species richness is the foremost determinant of the geographic distribution of parasite diversity.

The biogeography of freshwater mussels of the family Unionidae provides an excellent illustration of this phenomenon. The larvae (glochidia) of these mussels are obligate external parasites on the gills or fins of fish, whereas the adults live a benthic sedentary life typical of most bivalves. Given some host specificity on the part of the parasitic larvae, and an apparent lack of competition among mussel species (Vaughn 1997), we would expect that the more fish species present in a suitable aquatic habitat, the more mussel species can successfully exist in that habitat. In accordance with this prediction, Watters (1992) found a strong linear relationship between the number of fish per river system and the number of mussel species in those systems (Figure 6.1). At this spatial scale (considering entire drainage systems as biogeographic units), host species richness explains most of the variation in parasite species richness across rivers. On a finer scale, however, this relationship partially breaks down. Using sampling sites within a single river system as units of study, a positive relationship between fish species richness and mussel species richness still exists, but the scatter of points is roughly triangular rather than nicely linear (Figure 6.1). Within a river, the number of fish species using a particular location sets an upper limit to the richness of mussel species, but other environmental variables can further limit mussel species richness at any given locality (Vaughn and Taylor 2000). Thus, a locality with many fish species might also have many mussel species, but not necessarily so; whereas a location with few fish species can have only few mussel species. It is only at a scale where local processes become unimportant that the number of parasite species becomes directly proportional to the number of host species present within a given biogeographic unit. Deviations from this simple relationship would indicate the action of other forces, acting either to constrain parasite diversity or to increase it beyond what is predicted from host species richness. Essentially, the relationship between host species richness and parasite species richness provides a null model. In the following section, wherever possible we consider spatial variation in parasite species richness *per host species*, and not in absolute parasite species richness. Thus host species richness *per se* cannot explain the patterns discussed next, unless otherwise specified.

Latitudinal Gradients in Parasite Diversity

One of the best-documented large-scale biogeographic patterns is the increase in species richness with decreasing latitude (Rosenzweig 1995; Gaston and Blackburn 2000; Macpherson 2002; Willig et al. 2003). For most plant and animal taxa, species richness peaks in the tropics and decreases toward the poles. This pattern must be the outcome of higher rates of diversification at low latitudes (Cardillo 1999). Several theories for this phenomenon have been proposed, not necessarily mutually exclusive, and each championing different processes that are probably all acting in synergy. Among the most popular is the area hypothesis (Rosenzweig 1995; Rosenzweig and Sandlin 1997). Put simply, there are more species in the tropics because this is where we find the greatest geographic area. Considering terrestrial environments, the most extensive of the recognized biomes is indeed the tropics. Another theory is the energy hypothesis, which states that higher energy availability in an area provides more resources that can support more species (Turner et al. 1987; Currie 1991; Wright et al. 1993). This hypothesis predicts a strong link between climatic variables and species diversity (Acevedo and Currie 2003). A larger proportion of the solar radiation is

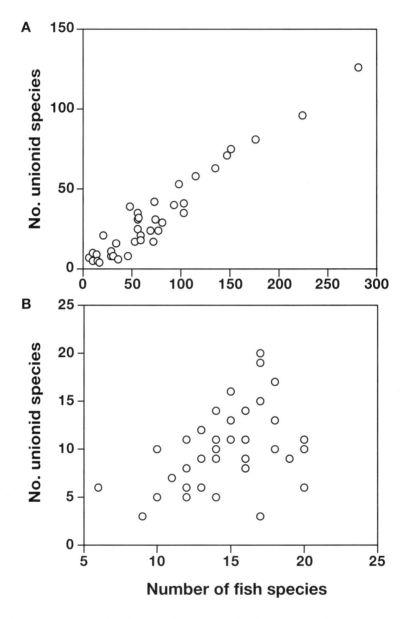

Figure 6.1. Relationship between the number of fish species and the number of unionid mussel species (A) across 37 riverine systems of the Ohio River drainage area in the central United States, and (B) across 36 sampling sites within a single riverine system, the Red River of the south-central United States. (Data from Watters 1992; Vaughn and Taylor 2000)

captured in the tropics than at higher latitudes, and this extra energy can be converted into additional biomass in the form of richer floras and faunas. We then have several versions of the time hypothesis, in which ecological time after a disturbance determines how many species can colonize or re-colonize an area, and the total evolutionary age of an area determines how much time has been available for species to evolve to fill all available niches (Pianka 1966; Rohde 1992). Tropical areas would need to be consistently older than temperate areas to explain latitudinal gradients in species richness. There is no consensus among ecologists regarding these and other mechanisms, probably because all of them are valid, at least to some extent and at certain scales.

What about parasites? Are there latitudinal gradients in parasite species richness, and if so what are the likely mechanisms behind these gradients? *A priori*, we might expect weak relationships between latitude and parasite species richness because the strength of these relationships increases with the average body size of the organisms considered (Hillebrand and Azovsky 2001), and parasites are generally small-bodied. The parasite groups for which biogeographic patterns are best known are the metazoan parasites of marine fishes (see Rohde 1993, 2002). Only a fraction of the faunas of marine fishes and their parasites has been studied, but it has been investigated in detail and a clear pattern has emerged: the diversity of monogenean parasites (and other groups of ectoparasites, to a lesser degree) increases with decreasing latitude or increasing water temperature (Rohde 1978, 1993, 2002; Rohde and Heap 1998). A simple explanation for these results would be that there are more parasite species in warmer tropical seas because they are inhabited by more host species. If cospeciation were the rule in the evolutionary history of host–parasite associations (Brooks and McLennan 1993b; Paterson et al. 1993; Hoberg et al. 1997), this would be a reasonable null hypothesis in all biogeographic studies of parasite diversity. This simple explanation, however, is not sufficient to account for the oceanic patterns reported by Rohde (1978, 1993, 1997, 2002). Using the number of parasite species *per host species examined* as a measure of relative diversity, parasite diversity is again higher in warm waters than in cold seas (Figure 6.2). The measure used in some sense controls for available area for parasites, and thus excludes the area hypothesis as a potential underlying mechanism.

The latitudinal gradient in the diversity of monogeneans and other ectoparasites of marine fishes could have several causes, some of which ap-

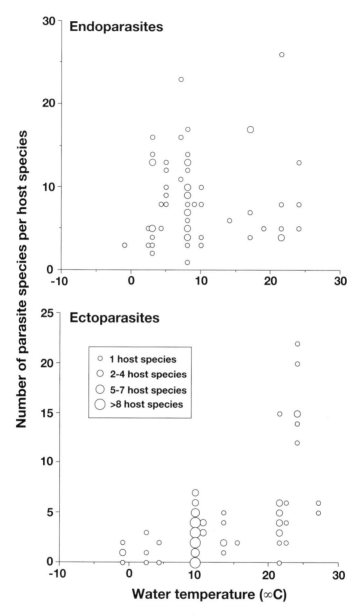

Figure 6.2. Relationship between the number of parasite species per host species and the water temperature at the sampling site. Results for endoparasitic helminths from 55 marine fish species (62 populations) and for ectoparasitic metazoans from 108 marine fish species (109 populations) are shown separately. (Data from Rohde and Heap 1998)

pear unlikely. For instance, the host specificity of monogeneans does not vary with latitude (Rohde 1978), so the greater relative diversity on tropical fishes is not simply the result of the same number of parasite species exploiting more host species. Also, the trend is not the product of unequal sampling effort or a phylogenetic artifact that could have resulted from fish lineages harboring diverse parasite faunas colonizing warm waters in the past: the effect of latitude or water temperature on parasite diversity remains after controlling for phylogenetic influences (Poulin and Rohde 1997; Rohde and Heap 1998). Other explanations exist, though, derived from the energy hypothesis, which states that species diversity is a function of the solar energy entering an ecosystem (e.g., Currie 1991; Wright et al. 1993), and from the evolutionary time hypothesis. One possibility is that taxa in warm waters experience higher rates of diversification because the shorter generation times and higher mutation rates resulting from higher temperatures lead to evolution itself proceeding faster (Rohde 1992, 1998b, 1999; see also Allen et al. 2002). The proposed link with generation times suggests a role for epidemiological parameters, as proposed in Chapter 3. Thus, according to Rohde (1992, 1997, 1999), over the same time period, and assuming a nonequilibrium state in which fish do not become saturated with parasite species, higher speciation rates in warm seas would lead to a greater parasite species diversity than in colder waters. The effective evolutionary time available for speciation in tropical waters is therefore greater than in temperate oceans. This hypothesis cannot account for the different latitudinal gradients in species diversity observed in ectoparasites and endoparasites of marine fishes (Figure 6.2). Intestinal helminths living in ectothermic fish hosts are also exposed to external water temperatures, but they have not diversified at a higher rate in the tropics. Rohde and Heap (1998) propose that other biological differences between internal and external fish parasites can explain the absence of latitudinal diversity gradients in endoparasites.

The temperature-mediated diversification is an interesting hypothesis that should apply equally well to all organisms, whether parasitic or not (Rohde 1992, 1999). It is, however, not the only explanation for the latitudinal gradient in the diversity of ectoparasites of marine fishes. Other comparative studies have shown that the body sizes of fish ectoparasites such as monogeneans decrease with decreasing latitudes (Poulin 1995a, 1996d). As discussed in Chapter 5, diversification may sometimes be greater in small-bodied taxa. Thus the greater diversity of fish ectoparasites in the tropics could result from greater diversification rates due to both environmental temperature and parasite body size; the two explanations are not mutually exclusive. This example illustrates the challenges impeding investigations into the causes of biogeographic patterns in parasite species diversity.

Another comprehensive investigation of latitudinal gradients in parasite species diversity has centered on helminth parasites of freshwater fishes (Choudhury and Dick 2000; Poulin 2001). The pattern observed, however, is not quite as expected. Choudhury and Dick (2000) found that temperate freshwater fish species are hosts to richer assemblages of helminth parasites than tropical freshwater fish species. Using the same dataset, Poulin (2001) eliminated the potentially confounding effects of uneven sampling effort, host body size, and host phylogeny to arrive at essentially the same conclusion: given the same sampling effort and for equal body sizes, temperate fish taxa harbor more parasite species than their tropical relatives (Figure 6.3). The only phylogenetic comparison between a temperate fish taxon and its tropical sister taxon that yielded a higher richness for the tropical taxon involved a contrast between congeneric eel species (*Anguilla* spp.), in which tropical species have remarkably richer parasite faunas than their temperate counterparts (Kennedy 1995; Marcogliese and Cone 1998). All other phylogenetic contrasts show higher parasite species richness in temperate fish species. At the moment, there is no explanation for this finding. It may be that temperate freshwater fishes possess features, such as those discussed in Chapter 4, that make them more likely to develop rich parasite faunas. For instance, species living at higher latitudes tend to have broader geographic ranges than those in the tropics (Stevens 1989), a factor that may promote the establishment of rich parasite assemblages.

In addition to the above studies on fish parasites, Cumming (2000) has carried out an investigation of latitudinal gradients in species richness of ticks in Africa. The spatial distribution records of ticks throughout the African continent are rather patchy. Given that the geographic ranges of ticks are better described by climatic variables than by host preferences, Cumming (2000) built logistic regression models in which climate data were used to predict the occurrence of each tick species among cells of a continent-wide grid. These were then used to derive "augmented" estimates of tick species richness in each cell and to show that tick species richness increases toward the equator. Unlike the studies of fish parasites described above, tick richness in Cumming's (2000) study is not corrected for host richness; it is not

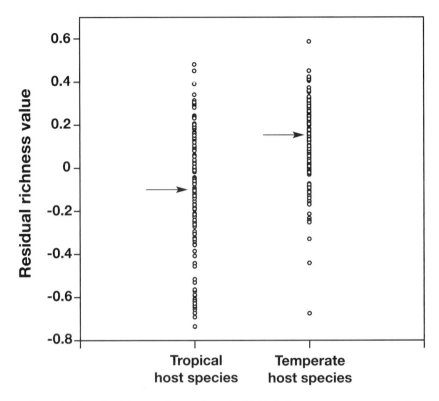

Figure 6.3. Species richness of gastrointestinal helminth component communities from tropical ($n = 125$) and temperate ($n = 129$) freshwater fish hosts. The values are residuals from a multiple regression on log-transformed data and are thus corrected for both host body size and sampling effort. Negative values indicate a component community with fewer species than expected based on host body size and the number of hosts examined, and positive values indicate the opposite. Arrows indicate the mean value for each group. (Data from Poulin 2001)

expressed on a "per host species" basis. However, latitudinal gradients in the richness of mammalian host species follow similar trends, and relative richness of ticks per host species may thus be constant across latitudes.

Other Gradients in Parasite Diversity

Other patterns are superimposed on latitudinal gradients in species richness. Studies on metazoan parasites of marine fishes have found that the diversity of parasitic monogeneans is greater in the Indo-Pacific Ocean than in the

Atlantic Ocean (Rohde 1980, 1986, 1993). This is a true difference, because it persists when parasite species richness is measured on a *per-fish-host* basis. Rohde (1986, 1993) rejected the area hypothesis as an explanation and instead proposed that the much greater age of the Pacific Ocean could account for its greater diversity in fish parasites. The degree of endemicity of the biotas of the two oceans suggests that exchanges of species between them have been limited and that their faunas developed independently, with richness proportional to time. An alternative hypothesis may be that during the last glaciation, the ice sheet covering the Atlantic was considerably larger than the one in the Pacific, causing higher extinction rates of parasites and other organisms in the Atlantic (Rohde 1993).

Among free-living organisms, longitudinal gradients in species richness are also known, although their explanations are more varied and less general than those behind latitudinal gradients (Gaston and Blackburn 2000). In the Pacific Ocean, the species richness (per fish species) of endoparasitic helminths of serranid fishes decreases from the Great Barrier Reef toward the Central Pacific area (Rigby et al. 1997). This trend parallels the decrease in the diversity of fishes and other reef organisms along the same axis. The center of diversity for coral reef fishes and their parasites appears to be the coastal waters of Indonesia and northeastern Australia. Rigby et al. (1997) suggest that as the distance from the center of diversity increases, oceanic islands and their associated reefs become smaller and more dispersed, making colonization by fishes or parasites more improbable. Simkova et al. (2001) also reported a longitudinal gradient in parasite species richness in central European freshwater fishes, with richness decreasing in an eastward direction. Their explanation is that the center of distribution of most of these fish species is in the west, and that fish expanding their range toward the east have gradually left parasite species behind. No doubt other longitudinal gradients in parasite diversity will have different explanations.

The final gradient considered here applies only to marine systems or to large lakes. The species richness of most major taxa of animals changes with increasing depth. For benthic invertebrates, species richness actually increases with depth down to a certain depth, beyond which it decreases again, although not all analyses support this widely accepted pattern (e.g., Gray 1994; Gage 1996). Whatever happens to the diversity of free-living animals in the deep-sea, their population density is often low. In particular, fish biomass is thought to decline with depth, although robust estimates are few (Merrett and Haedrich 1997). If potential host species occur at low population density in the deep-sea, we might expect that they would harbor a low species richness of parasites. For a parasite species to invade and successfully establish in a host population, host population density must be above a certain threshold (see Chapter 3). Indeed, most investigations of parasitism in deep-sea fish species indicate that parasite diversity is lower than in shallow-water fish species. This pattern is sometimes not apparent in mesopelagic fishes, which live off the continental slope at moderate depth and often migrate vertically to feed, but it is clear in benthic fishes (Noble 1973; Campbell et al. 1980; Gartner and Zwerner 1989; Campbell 1990; Bray et al. 1999; Marcogliese 2002). The existence of benthic fish species with a huge bathymetric range provides opportunities to sample populations of the same fish species at different depths to make comparisons of parasite species richness not confounded by host phylogeny. Such comparisons often provide telling evidence of a depth gradient in parasite abundance, paralleled by a decrease with depth in the abundance of free-living invertebrates that may serve as intermediate hosts for helminths (Figure 6.4); the decrease in abundance is mirrored by a decline in species richness. A phylogenetic analysis of deep-sea digeneans by Bray et al. (1999) has indicated that there have been few transitions from shallow to deep-sea habitats over the evolutionary history of digeneans, but that these few transitions have sometimes been followed by radiations. Indeed, some digenean species, like the aptly named *Profundivermis* sp., only infect their host fish in the deeper parts of its bathymetric range (Bray et al. 1999). Thus, despite its generally low diversity in parasites, the deep sea is still a place where new parasite species originate. There may even be hot-spots of parasite diversity in the deep sea: recent evidence suggests that parasite species richness may be relatively high in deep-sea hydrothermal vent communities, where high densities of hosts and warmer temperatures may facilitate parasite diversification (Moreira and López-Garcia 2003; de Buron and Morand 2004). Knowledge of parasitism in the deep sea is patchy at best and we will need much more information before we get a clearer picture of how diverse parasites in general are in the abyss and what roles they play down there.

Host Introductions and Parasite Species Richness

The processes discussed above and responsible for the current geographic distribution of parasite diversity all act slowly, over ecological or more likely

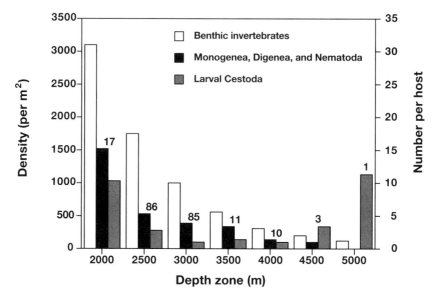

Figure 6.4. Density of benthic invertebrates and number of parasites per host (for all species combined) as a function of depth, for the fish *Coryphaenoides armatus* (Macrouridae) off the New York Bight. Numbers above the bars are the numbers of fish examined. (Data from Campbell et al. 1980)

evolutionary time scales. Whether involving the dispersal of species from a center of diversity or high speciation rates in certain areas, these processes shape parasite biodiversity slowly, and what we see today is the product of their past action. Recently, however, human activities have had substantial—and immediate—effects on parasite diversity. The most frequently mentioned effect is that of increased extinction rates (see Chapter 7). However, species introductions are also changing the face of parasite diversity in many areas of the globe. Deliberate introductions, particularly of vertebrates, were common over the past few centuries, whereas more recently, accidental introductions have increased dramatically with increased international trade and travel. For example, it has been estimated that, on any given day, ships used for maritime transport carry more than 3,000 different species of marine organisms in their ballast water (Carlton and Geller 1993). The establishment of exotic host species in new areas can have several effects on the distribution of parasite diversity among geographic areas and among host species as well.

First, when a few individuals of an exotic animal species are introduced into a new area, they are rarely accompanied by a full complement of the parasite species that exploited them in their original habitat. This is a simple consequence of the low prevalence and aggregated distribution of most metazoan parasites among their hosts. There is ample evidence that founder host populations establishing in new areas harbor species-poor assemblages of parasites compared with the original host population. Torchin et al. (2002, 2003) reviewed studies on the parasites of exotic marine and terrestrial invertebrates and vertebrates and found, on average, a threefold lower parasite species richness in the introduced range than in the native range. Similarly, possums *Trichosurus vulpecula* introduced in the nineteenth century from Australia to New Zealand left behind most of their parasite species: the parasite species richness of possums in New Zealand is much lower than in Australia (Stankiewicz et al. 1997a), and the parasite species of possums introduced from the New Zealand mainland to small offshore islands is lower still (Stankiewicz et al. 1997b). Introduced island populations of other host species also show lower parasite species richness than their mainland counterparts (e.g., Fromont et al. 2001; Pisanu et al. 2001; Goüy de Bellocq et al. 2002; Wilson and Durden 2003). The introduction of host species to new areas has thus created new host populations (i.e., new parasite component communities) with reduced parasite species richness. Because some parasite species can colonize exotic hosts, the reduced parasite species richness of introduced hosts is in part reversible. If this happens, however, the nature of the parasite assemblage will change. For instance, rainbow trout, *Oncorhynchus mykiss*, has been introduced from its native heartland in Kamchatka and western North America, to locations all over the world. The absolute and relative number of helminth species specific to salmonid fish found in parasite component communities of rainbow trout decreases with increasing distance from its heartland (Figure 6.5). However, in parts of North America and Europe, the total parasite species richness of introduced trout populations rivals those of the heartland after the colonization of the fish by local species of generalist parasites (Kennedy and Bush 1994). In other areas, such as Chile and New Zealand, even including the local generalist species of helminths that now exploit rainbow trout, total parasite species richness is only a fraction of what is observed in the native heartland (Kennedy and Bush 1994). Thus the introduction and establishment of nonindigenous host populations results in the formation of

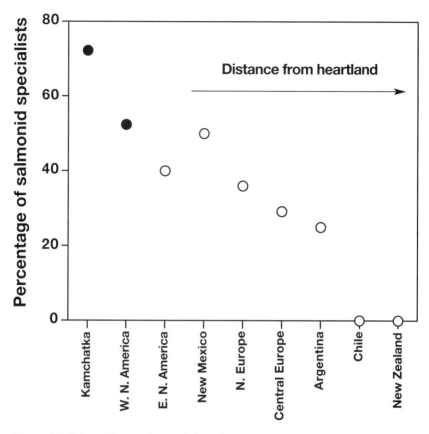

Figure 6.5. Salmonid specialists in helminth component communities of the rainbow trout, *Oncorhynchus mykiss,* relative to the distance from the host species' heartland. Rainbow trout originated in the Pacific region of North America and Kamchatka (black circles). (Data from Kennedy and Bush 1994)

parasite component communities with altered composition and almost invariably lower species richness.

Second, if they are not too host-specific, the few parasite species that accompany introduced hosts can infect indigenous host species and change the species richness of their parasite assemblages. For instance, the parasitic copepod *Lernaea cyprinacea,* introduced worldwide with the carp *Cyprinus carpio* and its other cyprinid hosts, had been recorded on more than 100 host species from 16 orders two decades ago (Kabata 1979); the list of host species with a parasite fauna now including this copepod has since

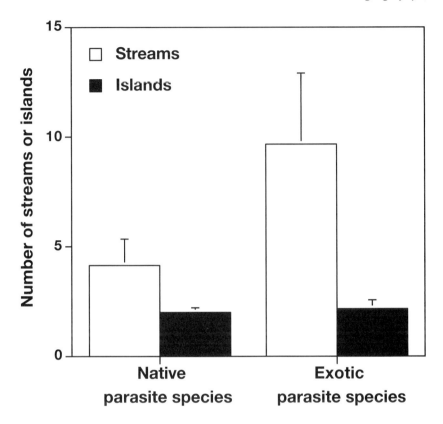

Figure 6.6. Number of streams and islands (mean ± standard error) where native and exotic fish parasites occur in the Hawaiian islands. Freshwater fishes were sampled from a total of 33 streams from four islands (Hawai'i, Kaua'i, Maui, and O'ahu). The data are for the six native and seven exotic species of metazoan parasites found on Hawaiian freshwater fishes. (Data from Font 1998)

grown much longer. Further, consider the freshwater fish fauna of the Hawaiian islands. It includes only five native species of stream fishes (four gobiids and one eleotrid) and three species of exotic poeciliid fish now widespread throughout the islands (Font and Tate 1994; Font 1998, 2003). Of the 13 species of helminths that parasitize these fish, seven are native and the remaining six were introduced along with the exotic fish. On average, the introduced parasite species are now found in twice as many streams among the islands than the native parasite species (Figure 6.6). This may be due to their tolerance of a wide range of environmental conditions and their

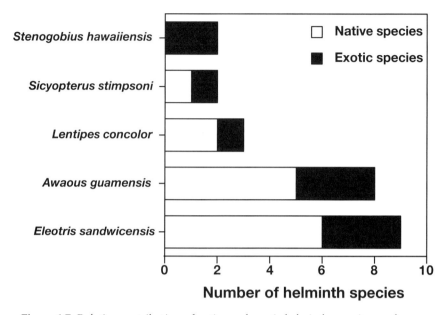

Figure 6.7. Relative contribution of native and exotic helminth parasites to the richness of the parasite fauna of the five freshwater fish species native to the Hawaiian islands. The upper four species are gobiids, and the lowest is an eleotrid. (Data from Font 1998)

dispersal abilities. Because of the low host specificity of three of the introduced parasites, the species richness of the parasite faunas of native fishes has now been increased (Figure 6.7), and the parasite fauna of one of the five native fish species apparently consists only of exotic parasites (Font 1998, 2003). The same scenario is unfolding elsewhere. In Australia, for instance, the exotic tapeworm *Bothriocephalus acheilognathi,* also one of the species spreading in Hawaii, is now found in all native fish species for which at least 30 individuals were examined and which occur in sympatry with the introduced original host species (Dove 1998; Dove and Fletcher 2000). The threat of other parasites recently introduced into Australian freshwater habitats also colonizing native species is very real (Dove 1998). Exotic parasites could replace native ones in the parasite faunas of certain hosts, but it appears more likely that they will simply cause the faunal species richness to increase. These recent developments can mask the effects of ecological and evolutionary processes that act much more slowly and can generate background noise in comparative analyses, such as those described in Chapters 3 and 4, that seek to uncover the determinants of parasite species richness in a group of host species.

Species introductions can have other effects on parasite species richness. Parasites may be introduced to new areas without their original hosts, although this is likely to be rare (Torchin et al. 2002). In any event, all these recent movements of hosts over large distances and short periods of time are rapidly changing the biogeography of parasite biodiversity and becoming a factor potentially more important than the natural processes discussed earlier.

Conclusions

In previous chapters, we saw that parasite species richness is not divided uniformly among host taxa or among parasite taxa: some host taxa harbor more parasite species than others, and some parasite taxa are much more speciose than others. We also saw that properties of hosts or parasites can promote the diversification of parasite assemblages. In this chapter, we find that the same conclusions apply to geographic areas. There are more parasite species per host species in certain geographic areas than in others (tropics versus temperate zones, shallow versus deep waters), and properties of these areas may have favored the speciation of parasite lineages or the accumulation of parasite species via colonization from other areas. Together, features of hosts, parasites, and geographic areas determine the extent and distribution of parasite diversity. As we saw in previous chapters, however, studies on the ecology and biogeography of parasite diversity are still few, and their results are not always in agreement. These early studies have set the stage for the ones to follow, and we can look forward to the emergence of more robust and general geographic patterns once additional investigations are available.

7

Parasite Extinctions

At the same time as we deplore the widespread extinction of plant and animal life on the planet, we are engaged in a battle against a small group of parasites for which we see only one satisfactory outcome: victory for us and the complete extermination of these parasites. The enemy in this battle consists of all parasites of humans and domestic animals, and the weapons include vaccines, drugs, and various barriers to infection, to name a few. Eradication of these parasites is proving extremely difficult and, in the end, reducing their abundance and thus their harmful effects may be a more reasonable objective. It is still ironic to strive for the extinction of a select group of organisms while so many efforts now go toward preserving biodiversity in general.

To be honest, parasites of humans and domestic animals such as sheep and cattle represent only a small fraction of total parasite biodiversity, and it is what is happening to parasite biodiversity as a whole that is of concern here. Alarms have been raised warning of the potential effects of human activities on parasite extinctions, especially via the fragmentation and general decline of numerous host populations (Rózsa 1992; Sprent 1992). In this chapter, we will review the characteristics of parasites that either make them liable to extinction or, conversely, provide them with hedges against extinction. We will then try to estimate the rates at which parasite species are currently being lost, to provide an idea of how serious the loss of parasite biodiversity is compared with the loss of other organisms.

Parasite Features That Lower the Risk of Extinction

Not all organisms face the same risk of going extinct: the possession of certain characteristics has apparently favored the survival of some taxa over others during the course of evolution (e.g., Jablonski 1986). What about parasites? Various features of parasites are often invoked as adaptations facilitating transmission success (see Poulin 1998a), but most of these are probably useless against the possibility of extinction. For instance, the extremely high fecundity of many parasites facilitates the completion of the life cycle under normal conditions, but when the host species has practically gone extinct, it does not matter how many eggs are produced by the remaining adult parasites: they may be doomed to extinction, too.

Bush and Kennedy (1994) suggested that other features of parasites act as hedges against local extinction of parasites within host populations or given geographic areas. In particular, they argue that low host specificity (i.e., the ability to exploit and survive in several host species at any stage in the life cycle) is a major factor favoring the persistence of parasites even after the local extinction of a host species (Figure 7.1). Clearly, the existence of a parasite species capable of surviving in only one host species is tightly associated with that of the host species: if the host disappears, so does the parasite. Low host specificity weakens the link between a parasite and the survival of any of its host species. In a parasite species with a two-host life cycle, for instance, low host specificity for *either* the intermediate or the definitive host would reduce the probability of extinction, and low host specificity for *both* intermediate and definitive hosts would reduce it even further. Every time a new intermediate or definitive host species is colonized by the parasite, via host switching, and added to the spectrum of host species exploited, the probability of extinction decreases. In Figure 7.1, we see that parasites that add new hosts to their repertoire shift their position from dark-shaded areas toward light-shaded ones, along either the *x*-axis or the *y*-axis, or both. Parasites with direct one-host life cycles can move along only one axis, whereas parasites with three-host cycles can move along a third axis as well.

Most basic textbooks of parasitology emphasize the high specificity of parasites for their hosts, which would suggest that most of them face a risk of extinction at least as high as that of their hosts. In fact, host specificity varies widely among related parasite species (Poulin 1992a). For instance,

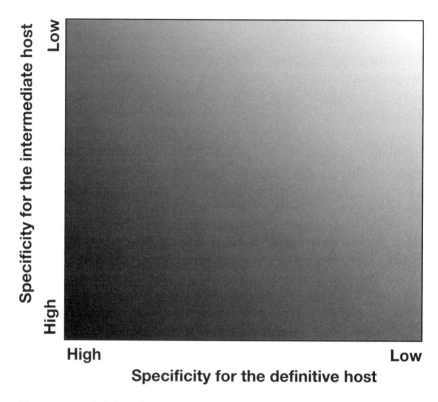

Figure 7.1. Probability of extinction of a hypothetical parasite with a two-host life cycle as a function of its specificity for both the intermediate and definitive host: the darker the shading, the higher the probability of extinction. (Based on ideas of Bush and Kennedy 1994)

a recent survey indicates that about three-quarters of the more than 600 known parasites of livestock are capable of infecting other host species, many of which belong to different mammalian orders (Cleaveland et al. 2001). So, for many parasites, there may be the possibility of surviving on alternative hosts should one host species disappear.

Bush and Kennedy (1994) also suggest that high prevalence in at least one if not all host species would also provide some security against extinction. Clearly, if members of a parasite species are found in, say, only 4% of the host individuals in a population, then they are at the mercy of stochastic events that often cause mass mortalities in animal populations; if the few hosts that they infect succumb to a bad winter, the parasites go locally extinct. The extinction probability spectrum of Figure 7.1 assumes a fixed average prevalence of infection. A parasite with relatively high host specificity may not face a high risk of extinction even if its host population declines, as long as it normally occurs in a high proportion of these hosts.

The situation may in fact be more complex, as there may be various trade-offs between traits like host specificity and average prevalence of infection. Adaptations to evade or combat host immune defenses must be costly; parasite species specializing on a single host species may achieve higher prevalence or intensities of infection in this host than if they had to invest in a wider range of evasive mechanisms to cope with a broad array of host species. We might thus expect a negative interspecific relationship between the number of hosts exploited and average prevalence, at any stage in the life cycle. Such a relationship was indeed found among a large number of metazoan parasite species of freshwater fish hosts (Poulin 1998c). It may be that few parasite species display both low host specificity and generally high prevalence of infection among these hosts. If rare in a particular host species, a parasite is likely to occur in other hosts, but if common in a host species it may occur in few or no other. Local extinction may be more likely for the former, as its main local host species can disappear, whereas global extinction may be more common for the latter, happening if its main or only host goes extinct. This trade-off between the number of host species used and average prevalence may not be universal: among a large sample of helminth species parasitic on birds, a *positive* relationship was found between the two variables (Poulin 1999b), suggesting that low host specificity and high prevalence can go hand in hand. Many factors can explain the different results obtained for bird and fish parasites (Poulin 1999b). The key point, however, is that in helminth parasites in birds, the combined action of low host specificity and high prevalence can buffer many species against extinction even after the disappearance of a host species. At the same time, for helminth species characterized by both high host specificity and low prevalence values, extinction may be a real risk.

In addition to host specificity, the complexity of a parasite's life cycle also determines their dependence on a set number of host species. Many helminths go through a succession of developmental stages in different host species, and therefore require the availability of a sequence of suitable hosts. Certain trematodes require up to four hosts to complete their life cycle. It follows that should any of these hosts disappear, the parasite could go extinct, regardless of the availability of other hosts. Recent evidence has challenged the view that the life cycle of helminth parasites is inflexible. For example, like all other members of its family, the nematode *Camallanus cotti* has a two-host life cycle: adult worms live in freshwater fish from which they release larvae, which then infect planktonic copepods as intermediate hosts. However, captive fish populations maintained in closed aquarium systems for several generations remain infected with the nematode, despite the complete absence of copepods in the water (Levsen 2001; Levsen and Jakobsen 2002). Instead of going extinct in these aquarium fish populations, the parasite has rapidly switched from a two-host to a one-host life cycle, adopting direct fish-to-fish transmission as soon as copepods became unavailable. Similarly, facultative truncation of the usual three-host life cycle is common among trematodes (Poulin and Cribb 2002). Trematodes in many families appear to have the latent ability to abbreviate their normal life cycle by skipping one transmission event—they can drop one host from the cycle when this host is locally unavailable. Thus many parasites could survive the complete disappearance of a host that normally plays an important role in their transmission and developmental cycle. Nevertheless, for most helminth parasites, a complex life cycle may be a liability when the abundance of one or more host species begins to decline.

Finally, parasite body size itself may be associated with the risk of extinction. Across nematode species parasitic on terrestrial mammals, the relationship between the intensity of infection per host mass and nematode body size follows a polygonal pattern similar to that observed in the density–body mass relationships of free-living animals (Morand and Poulin 2002). There seems to be a peak in parasite intensity associated with some intermediate body size; intensity decreases as body size moves either way from this optimal value (Figure 7.2). Most living nematode species tend to have body sizes close to this optimum (Morand and Poulin 2002). The lower side of the polygon may correspond to the threshold below which nematode populations cannot persist—a threshold below which the R_0 value resulting from the combination of parasite traits does not permit the maintenance of the parasite in the host population. Extinction may thus be a real risk for species outside the polygon, that is, for nematodes with very small or very large body sizes (Figure 7.2). This prediction has not been tested; it is merely a speculation derived from the observed relationship between intensity of infection and parasite body size.

ecessary for parasite transmission ($R_0 < 1$), the parasite will dis-
apter 3). The reliance of parasites on their hosts makes host
decline or extinction the main cause of parasite extinctions.
he reasons behind the extinction of hosts (free-living species) are
most parasite extinctions. These reasons have been previously
isk et al. 1994; Sutherland 1998), and here it suffices to say that
host species goes extinct, one or more parasite species most
ccompany it into oblivion.

inths with complex life cycles, the disappearance of only one of
ed hosts may be sufficient for either their local or global disap-
some lakes in northern Michigan, the number of species of lar-
des using the snail *Stagnicola emarginata* as intermediate host
d by more than 50% over a period of just over 50 years, with
ies now found in the area (Cort et al. 1937; Keas and Blanke-
'). Increasing human activity around the lakes—and its associ-
on the abundance of the aquatic birds used as definitive hosts
atodes—is the most likely cause for the decrease in parasite di-
s and Blankespoor 1997).

can also go extinct independently of their hosts. Environmen-
may affect parasites more severely than they affect hosts, caus-
ppearance of the former but not the latter. Chemical pollution,
, is now a ubiquitous consequence of human activities, especially
in aquatic environments. There is mounting evidence that it has
ts on the parasites of aquatic animals, even at levels where the
selves are not affected (Khan and Thulin 1991; Poulin 1992b;
'97). The free-swimming infective and dispersal stages of many
e susceptible to toxic chemicals, to the extent that none may sur-
certain conditions. The influence of pollution can result in dif-
parasite species richness among localities or host populations. For
arcogliese and Cone (1996) have found that parasite species rich-
, *Anguilla rostrata*, decreased as water pH values declined (Fig-
this case, some parasites like monogeneans were directly affected
cipitation, whereas others were only indirectly affected because
r intermediate hosts could not survive in acidic waters. This was
trematodes, whose snail intermediate hosts were absent from sites
pH (Marcogliese and Cone 1996). The direct effects of different
llutants on different parasite taxa are summarized in Table 7.1;

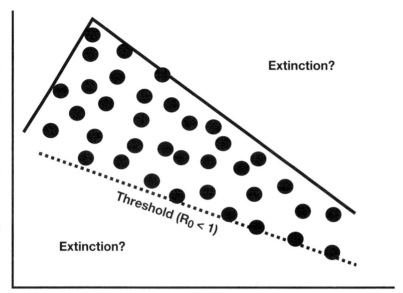

Figure 7.2. Idealized relationship between the intensity of parasite infection (per host mass) and parasite body size, across species of nematodes parasitic on terrestrial mammals. Few points fall outside the polygon, suggesting that parasites in those areas of the parameter space are at high risk of extinction. (Data from Morand and Poulin 2002)

Causes of Parasite Extinction

The tight association between parasites and their hosts means that the fate of the former is strongly coupled with that of the latter. As discussed earlier, low host specificity can weaken this coupling, but to some extent the extinction of a host species is always likely to have negative consequences for its parasites. A significant proportion of extant parasite species have originated via cospeciation with their hosts (see Chapter 4); similarly, the demise of a host species can mean the coextinction of its parasite species (Stork and Lyal 1993). Many highly host-specific parasite species, such as lice on terrestrial vertebrates and monogeneans on aquatic vertebrates typically found on a single host species, will invariably disappear if and when their host goes extinct. Actually, the complete extinction of the host is not necessary for parasite extinction; once the host population density drops below the threshold necessary for parasite transmission ($R_0 < 1$), the parasite will disappear (Chapter 3). The reliance of parasites on their hosts makes host population decline or extinction the main cause of parasite extinctions. Therefore, the reasons behind the extinction of hosts (free-living species) are also driving most parasite extinctions. These reasons have been previously reviewed (Sisk et al. 1994; Sutherland 1998), and here it suffices to say that each time a host species goes extinct, one or more parasite species most likely will accompany it into oblivion.

For helminths with complex life cycles, the disappearance of only one of their required hosts may be sufficient for either their local or global disappearance. In some lakes in northern Michigan, the number of species of larval trematodes using the snail *Stagnicola emarginata* as intermediate host has declined by more than 50% over a period of just over 50 years, with only 8 species now found in the area (Cort et al. 1937; Keas and Blankespoor 1997). Increasing human activity around the lakes—and its associated effects on the abundance of the aquatic birds used as definitive hosts by the trematodes—is the most likely cause for the decrease in parasite diversity (Keas and Blankespoor 1997).

Parasites can also go extinct independently of their hosts. Environmental changes may affect parasites more severely than they affect hosts, causing the disappearance of the former but not the latter. Chemical pollution, for example, is now a ubiquitous consequence of human activities, especially noticeable in aquatic environments. There is mounting evidence that it has direct effects on the parasites of aquatic animals, even at levels where the hosts themselves are not affected (Khan and Thulin 1991; Poulin 1992b; Lafferty 1997). The free-swimming infective and dispersal stages of many parasites are susceptible to toxic chemicals, to the extent that none may survive under certain conditions. The influence of pollution can result in differences in parasite species richness among localities or host populations. For instance, Marcogliese and Cone (1996) have found that parasite species richness in eels, *Anguilla rostrata,* decreased as water pH values declined (Figure 7.3). In this case, some parasites like monogeneans were directly affected by acid precipitation, whereas others were only indirectly affected because one of their intermediate hosts could not survive in acidic waters. This was the case of trematodes, whose snail intermediate hosts were absent from sites with a low pH (Marcogliese and Cone 1996). The direct effects of different types of pollutants on different parasite taxa are summarized in Table 7.1;

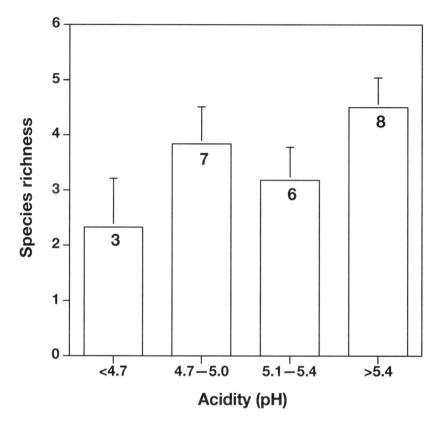

Figure 7.3. Species richness (mean ± standard error) of metazoan parasites of eels, *Anguilla rostrata,* from localities in Nova Scotia with different levels of acidity. Data are from the 24 localities where at least six fish were surveyed for parasites. The numbers in the bars are the number of localities in each acidity class. (Data from Marcogliese and Cone 1996)

although generally negative, the effects are variable. In contrast, the effects of other types of pollution, such as eutrophication, are generally positive (Lafferty 1997). Clearly, the overall influence of pollution on parasite diversity is not easily predictable, but it is also not negligible.

Other environmental changes can affect parasites and possibly cause the extinctions of a few or many species. Again, too little is known about the interactions of parasites and environmental factors to make safe predictions. Global warming, for instance, if moderate, can actually benefit the proliferation of many types of parasites (Marcogliese 2001b); however, the pro-

Table 7.1

Summary of the direct effects of different types of pollution on helminth parasites (from Lafferty 1997)

Parasite Taxon	Pulp Mill Effluent	Crude Oil	Industrial Effluent	Sewage Sludge	Acid Rain	Heavy Metals
Monogenea	–/+	++	+/–	0	–	?
Trematoda	–/+	– –	–/+	–/+	–	– –
Cestoda	?	?	–/+	+	–	– –
Acanthocephala	?	–	– –	– –	+	– –
Nematoda	+/–	+	–/+	0	?	?

Symbols: + and – = positive and negative effects based on single studies; ++ and – – = positive and negative effects based on many consistent studies; +/– = a positive bias among many inconsistent results; –/+ = a negative bias among many inconsistent results; 0 = no effect; ? = no studies are available.

liferation of a few species could drive others to extinction. Although it is obvious that when hosts go extinct, so too do parasites, we know next to nothing about the other factors that may cause parasite extinctions.

Dynamics of Local Parasite Extinctions

On small spatial scales, parasite extinctions are common and trivial. For instance, many parasites exist as metapopulations (*sensu* Hanski and Simberloff 1997) in patchy habitats (fragmented host populations). Stochastic processes can lead to the extinction of the parasite in one fragment, but because of the connectivity among fragments, the unparasitized patch is rapidly recolonized by parasites from adjacent patches.

The spatial mosaic of presences and absences of certain parasite species among host populations allows the dynamics of these local extinctions to be visualized. The parasite species found in a given host population (or habitat fragment, or island) are not a random sample of the species available from the regional species pool. One commonly observed pattern of parasite species distributions among host populations is the nested subset pattern (Patterson and Atmar 1986; Worthen 1996; Worthen and Rohde 1996; Wright et al. 1998). This pattern corresponds to the situation where the parasite species from a host population harboring few parasite species are consistently nonrandom subsets of host populations with progressively richer parasite assemblages. Perfect nestedness, or its opposite antinestedness, is extremely rare (Jonsson 2001); instead, the presence/absence matrix repre-

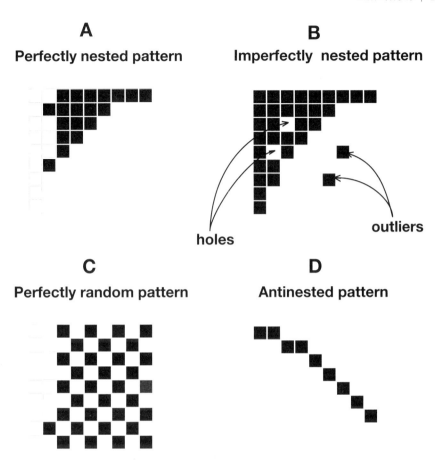

A
Perfectly nested pattern

B
Imperfectly nested pattern

holes

outliers

C
Perfectly random pattern

D
Antinested pattern

Figure 7.4. Matrices of presence/absence for several parasite species among several host populations. Each row represents a parasite species, each column represents a host population (or habitat fragment), and a black square indicates the presence of a parasite species in a host population. The patterns shown are a perfect nested pattern (A), an imperfect nested pattern, with outliers and holes (B), a random checkerboard pattern (C) and an antinested pattern (D).

senting the distributions of parasite species among host populations is more often somewhere between an imperfect nested pattern and a random one (Figure 7.4). In an imperfectly nested matrix, outliers may indicate parasite species with high dispersal or colonization potential, but with low probabilities of persistence. Similarly, holes in imperfectly nested matrices (Figure 7.4) may be associated with parasite species prone to local extinction.

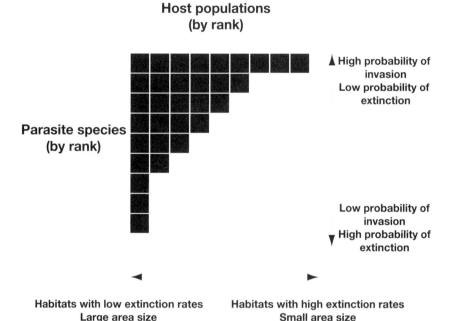

Figure 7.5. Nested matrix of presence/absence for several parasite species among several host populations (as in Figure 7.4). Host populations are ranked from those with low extinction rates and high parasite diversity (left) to those with high extinction rates and low parasite diversity (right). Parasite species are ranked from those with high probabilities of invasion and low probabilities of extinction (top) to those with low probabilities of invasion and high probabilities of extinction (bottom).

Nested patterns can be used to determine whether extinction probabilities are associated with various features of parasites. For this, we must make the reasonable assumptions that nested matrices are the product of colonization and extinction processes and not of interspecific interactions among parasite species, and that probabilities of colonizing a host population and of subsequent local extinction are attributes of parasite species (Morand et al. 2002). If host populations are ranked from left to right in the matrix in order of decreasing parasite species richness, and parasite species are ranked from top to bottom in decreasing order of frequency of occurrence in host populations, we obtain a matrix, likely to approximate a nested subset pattern, in which host populations and parasite species are arranged by extinction probabilities (Figure 7.5). In other words, parasite species near the top of the matrix have intrinsic properties giving them high probabilities of invading and colonizing host populations and low probabilities of extinction. Those at the bottom of the matrix have properties associated with low probabilities of colonization and high probabilities of extinction (Goüy de Bellocq et al. 2003). These properties may include host specificity or complexity of the life cycle. It is then possible to correlate these properties with the species' rank in the matrix to test whether any feature of parasite species is linked with high risks of local extinction. This approach could shed some light on the determinants of parasite extinctions at small spatial and temporal scales.

These small scales are relevant to parasite population processes. The trematode *Schistosoma mansoni* and its rat definitive hosts in Guadeloupe provide a good illustration of the spatial and temporal dynamics of parasite extinctions (Théron et al. 1992). The aquatic transmission sites, where rats acquire the parasite via free-swimming cercariae released from snail first intermediate hosts, are patchily distributed within swampy forest habitats. From year to year, extinction of the parasite occurs in some sites, followed by repopulation from adjacent sites in subsequent years. The presence and abundance of the parasite within any locality are unstable over time, whereas at the metapopulation level they are stable (Théron et al. 1992).

Estimated Rate of Global Parasite Extinctions

The types of extinctions discussed in the previous section are merely the outcomes of demographic processes acting over short ecological timescales (Bush and Kennedy 1994). Unless they stop being reversible, they are not evolutionary extinctions in the "extinction is forever" sense.

True extinctions of whole parasite taxa also happen. Because most soft-bodied parasites such as helminths rarely leave fossil remains, it is difficult to compare the parasite faunas of past eras with today's extant parasite diversity. The common occurrence of fossil insects caught in amber along with their parasites, however, has made it clear that some parasite species, or even higher taxa, that existed millions of years ago, are not represented today (e.g., Poinar et al. 1997; Poinar 1999). We can only assume that past rates of parasite extinctions have been similar to those of free-living organisms, with mass extinction events punctuating long periods with low rates of

Table 7.2

Estimated numbers of parasitic trematodes, cestodes, nematodes, and acanthocephalans at risk of extinction

	Chondrichthyes	Osteichthyes	Amphibia	Reptilia	Aves	Mammalia	Total
Percentage of host species listed as threatened[a]	2	2	2	3	11	11	
Number of parasite species at risk							
Trematoda	1	117	23	113	1,085	409	1,748
Cestoda	27	89	6	33	1,546	510	2,211
Acanthocephala	—	25	3	6	86	33	153
Nematoda	3	53	53	192	1,007	328	1,636

[a]From Smith et al. 1993a.

background extinctions. In the past century or two, the number of documented species extinctions has risen sharply due to the expansion of human
activities (Smith et al. 1993a), and some have suggested that, given current
extinction rates, half of the living bird and mammal species will have dis-
appeared within 200 to 300 years (Smith et al. 1993b). Of course, such estimates are plagued by uncertainties (Heywood et al. 1994) and perhaps
they can best be seen as worst-case scenarios. Still, getting an approximate
value for extinction rates can serve to mobilize efforts for the preservation
of biodiversity.

In this context, can we estimate current rates of parasite extinctions?
Using the proportions of threatened species in the major groups of vertebrates reported by Smith et al. (1993a) and the estimates of parasite species
richness presented in Table 2.1, we can at least estimate what percentage
of parasitic helminths are currently under threat of extinction. Host specificity matters, here, because host extinction may not be followed by parasite extinction if all parasites utilize other hosts. For instance, the known
parasites of the endangered Florida panther, *Felis concolor coryi,* are all
found in other, more common host species (Forrester et al. 1985): if the panther goes extinct, its parasites will not. Nevertheless, even when taking host
specificity into account (probability of extinction divided by the mean number of host species used by a parasite), we still find that several thousand
parasite species are currently threatened, as detailed in Table 7.2. These estimates represent 6 to 8% of the global species diversity of the major
helminth taxa, a figure that is likely to climb further if conservation policies
aimed at preserving vertebrates are unsuccessful (Smith et al. 1993a,
1993b).

The estimates in Table 7.2 are based on global averages of the proportions of threatened host species in the major groups of vertebrates; these values actually vary widely among geographic regions (Smith et al. 1993a).
Our estimates are also clearly too low, for at least two reasons. First, they
consider only how the fate of vertebrate definitive hosts influence the probability of helminth extinction: they rest on the assumption that intermediate
hosts remain available. In reality, an abundant supply of definitive hosts is
only one requirement for parasite existence; if an essential intermediate host
were to disappear, the parasite would follow no matter how abundant its
definitive hosts are. Most helminths use invertebrates as intermediate hosts
and extinction rates of invertebrates are even more uncertain than those of

vertebrates (Smith et al. 1993a, 1993b; Heywood et al. 1994). Therefore it is not possible to guess how severely the numbers in Table 7.2 underestimate the extinction threat faced by parasitic helminths. Second, as discussed earlier, parasites can go extinct independently from their hosts and, in addition to host-linked extinctions, there must be others that are left out of our estimates. This suggests that actual extinction rates of parasites may be higher than those of free-living animals. When a host species goes extinct, at least some of its parasite species may accompany it, but when a parasite species goes extinct for other reasons, its hosts will not follow it.

Conclusions

When parasites go extinct, they do so quietly, without fanfare. Most parasite extinctions probably go completely unnoticed, and if they were noticed they would most likely be greeted by cheers. Yet, they represent a considerable proportion of extant biodiversity and, as will be seen in the next chapter, their influence is not always negative. As their important roles in evolution and ecosystems become clearer, we will need to gather the necessary information to forecast their extinctions and their consequences and, in some cases, it may even prove beneficial to attempt to prevent them. In the meantime, like so many other invertebrates, many of them will continue to disappear before scientists have even described them.

8

Parasite Diversity Driving
Host Evolution

In earlier chapters, we have seen how host characteristics may have driven the evolution of parasite diversity. In turn, parasites may affect the evolution of certain host traits and be responsible for the maintenance of genetic and phenotypic variation within host species, as well as the evolution of biodiversity on a larger scale. In this chapter, we explore how parasitism, and parasite diversity specifically, may influence host evolution on several levels.

Parasites fit the criteria necessary to be direct agents of natural selection (Goater and Holmes 1997). Both the number of parasites of given species, and the number of parasite species, vary among individual hosts in a population, and this variation is often coupled with variation in host fitness. Phenotypes associated with lower infection levels, or capable of achieving high fitness despite infection, will therefore be favored by selection. The importance of parasites as agents of selection becomes clear when their numbers reach a level at which they noticeably affect host abundance. Numerous studies have shown that parasites and pathogens have the ability to regulate the population dynamics of their hosts (Scott 1987; Grenfell and Dobson 1995; Hudson et al. 1998). Because parasites can determine which individual hosts die and which survive, parasitism may be seen as a major evolutionary driving force (Hochberg et al. 1992; Goater and Holmes 1997; Summers et al. 2003).

Traits associated with resistance to infection are the ones under direct selection by parasites. The main mechanism of resistance against parasites is the immune system. It is not the only way animals can resist or avoid infection; a broad panoply of morphological and behavioral adaptations can also serve the same purpose. For instance, animals can recognize and avoid oviposition sites where their offspring would be at risk of parasitic infection (Lowenberger and Rau 1994; Kiesecker and Skelly 2000); site-selection strategies may have evolved as antiparasite defenses just as the immune system did. No defense mechanism, however, is as widespread and as broadly efficient as the immune system. Because it is but one of the integrated components of the whole physiology of host organisms, and because it requires some investment, we can assume that any increase in this investment is made at the expense of other functions. Thus, selective pressures from parasites will determine the optimal investment in immune defense for any given host species. Theoretical frameworks for the evolution of life-history strategies are based on trade-offs (Stearns and Koella 1986; Roff 1992; Stearns 1992; Charnov 1993). In this light, parasitism becomes a major driver of host evolution: the genotypes/phenotypes favored in the presence of parasites are different from those that would be selected if parasites were absent from a host population.

Most studies on this topic have only considered the potential effect of single parasite species on the evolution of particular host traits (e.g., Lafferty 1993; Koella 2000). We can hypothesize that the influence of parasitism should increase with the diversity of parasites a given host must face. Thus parasite species richness may be a good predictor of the extent to which the evolution of host life-history traits (Morand 2000) or immunity (Rigby and Moret 2000; Dupas et al. 2004) has been shaped by parasitism. In this chapter, we review briefly how host life-history strategies and immune defenses may be influenced by parasite diversity. Our review is meant only as an overview and not as a comprehensive survey of the work in this active research area. The reader is encouraged to consult the literature cited for more information. Our aim is simply to illustrate some of the wider ramifications of parasite biodiversity. With this in mind, we conclude the chapter with a discussion of the potential for parasites to act as promoters of biodiversity by facilitating host speciation.

Parasites and Host Life-History Traits

Of all life-history parameters, the onset of maturity is one of the most important events in the life cycle of organisms. The maximization of lifetime fitness implies an optimal age for first reproduction (Roff 1992). Life-history models predict that increasing adversity with age should select for an earlier age at maturity (Figure 8.1). In accordance with this expectation, Lafferty (1993) showed that a marine snail, *Cerithidea californica*, matures at smaller sizes in localities where the level of parasitism by a castrating trematode is high. Variation in trematode prevalence among snail populations can also be associated with variability in other life-history traits, such as the rate of egg production (Krist 2001). Examples like these suggest that parasitism selects for life-history strategies that differ from those we would expect in the absence of parasitism.

Can parasite species richness act in the same manner? The combined effects of several parasite species could exert strong selective pressures on host life-history strategies. Consider the link between larval helminth species richness and age at maturity among fish species. We might expect that host species harboring a high number of larval helminth species should mature earlier, because larval helminths often require, and may even cause, the death of their hosts to complete their life cycle. Moreover, larval helminths can often live much longer than their intermediate fish hosts (e.g., the nematode *Anisakis* spp, Anderson 2000). A comparative study of larval helminth species richness across 33 fish species showed that fish harboring a high larval parasite species richness mature at an earlier age (Morand 2003; Figure 8.1). Species of fish that accumulate diverse species of larval helminths may experience a decrease in overall survival, and have thus been selected to reduce their age at maturity. It must be emphasized that selection on one host trait does not occur in a vacuum: linkages between various traits are numerous. For instance, any reduction in age at sexual maturity may be coupled with the maturation of the immune system and the senescence that affects ageing individuals (Figure 8.2). Senescence coincides with the onset of many parasitic and autoimmune diseases (Coe 2002), and there are thus links between life-history traits and the immune system (discussed later).

Host Metabolism and Parasites

Parasites induce a cost to their hosts because of pathogenicity, immune activation, and increased metabolic activity. Several studies have shown that parasite infection increases the basal metabolic rate of infected hosts. For example, the red grouse *Lagopus lagopus* increases its metabolic rate by 16% after infection with nematodes (Delahay et al. 1995). Immune challenges also cause increases in basal metabolic rate in birds (Svensson et al. 1998; Ots et al. 2001; Martin et al. 2003).

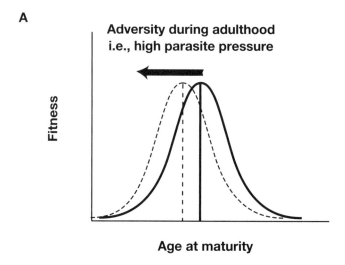

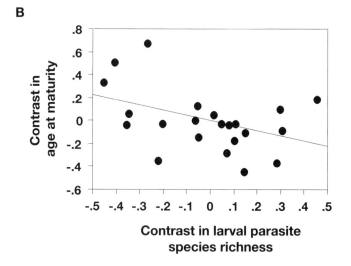

Figure 8.1. Relationships between host age at maturity, host fitness, and parasitism. (A) The optimal host age at maturity is the age at which host fitness, or lifetime reproductive success, is maximized. Increasing adversity during adulthood, such as that exerted by parasites, should select for earlier age at maturity. (B) Parasites that accumulate with host age, such as the larval stages of many helminths, should lead hosts to mature earlier. This is seen in a comparative test relating age at maturity and the richness of larval helminths in freshwater fish species; each point represents a phylogenetic contrast corrected for host sampling effort. (S. Morand, unpublished data)

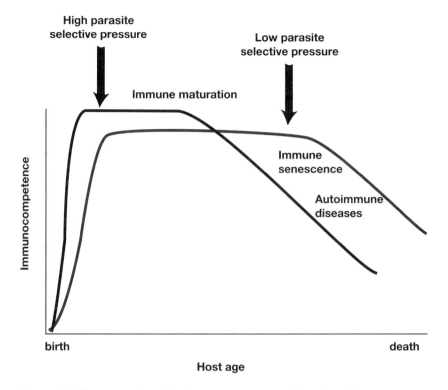

Figure 8.2. The maturation of the immune system may depend on the pressure exerted by parasites. It is characterized by a phase of relatively rapid improvement in immunocompetence, followed by a period during which the immune system functions at its peak, and finally by the senescence of the immune system, when parasites and autoimmune diseases become a serious risk. Greater parasite pressure should select for a decrease in age at sexual maturity, with consequences for the maturation of the immune system and the senescence that affects ageing individuals.

The basal metabolic rate, or BMR, is an indication of energy expenditure in animals (Koteja 1991), where it represents the minimum energetic cost of maintaining the basic life activity of an organism. The BMR has been related to a great number of variables across animal species, such as body mass, dietary habits, reproductive strategies, and phylogenetic relatedness (McNab 1980, 1992; Harvey et al. 1991). However, although there is a clear allometric relationship between metabolism and body size, much of the residual variation in the BMR remains to be explained (Elgar and Harvey 1987).

Could parasitism be one of the factors that selects for an animal's BMR? Morand and Harvey (2000) investigated the link between the BMR of mammals and the selective pressures exerted by parasites. They hypothesized that mammal species subjected to infection by numerous parasite species should have evolved higher BMR values than those exposed to few parasite species, to meet the costs of an efficient immune response. Because BMR and immune functions decline with age within any mammal species, these authors propose the second hypothesis that high parasite species richness should have a negative influence on host longevity. Morand and Harvey (2000) found that hosts suffering from high parasite species richness have both a higher BMR and a lower life expectancy than hosts exposed to a lower parasite diversity. An alternative hypothesis is that hosts with high BMR relative to their body size are more active and feed at higher rates and therefore have higher colonization rates by parasites because of their greater movement and foraging (Gregory et al. 1996). Because this hypothesis, like that of Morand and Harvey (2000), also predicts a positive relationship between BMR and parasite species richness, it is not yet clear whether high BMR values in certain mammal species are a consequence or a cause of high parasite species richness.

Hosts must invest in immune functions to control and to reduce the damages they may incur from parasitism. However, immune responses are energetically costly to the host (Lochmiller and Deerenberg 2000), with any energy allocated to immunity being lost to other functions. Accordingly, a recent study showed that BMR is higher in mice with suppressed production of lymphocytes T and B than in normal mice (Raberg et al. 2002). This suggests that adaptive immune responses may have evolved with a reduction in BMR. Indeed, Sorci and Morand (unpublished data) have shown that BMR in captive mammals is negatively correlated with the number of circulating lymphocytes in their blood. These results are difficult to reconcile with those of Morand and Harvey (2000) mentioned above; it may be that the cost of immune responses, and their relationship with BMR, depends on whether they are innate or acquired. Hence, the links between BMR, life-history traits, and immune defense are not clear and need further investigations.

Parasites and Host Investment in Immune Functions

For a host, an immune response represents an investment because it incurs costs measurable as reduced energy available for other functions (Owens

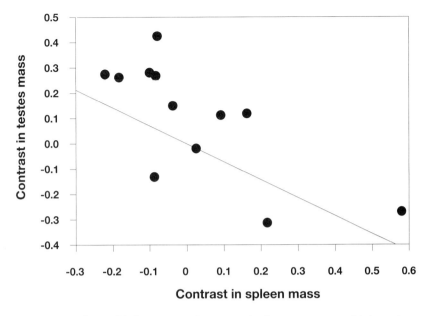

Figure 8.3. Relationship between testis mass and spleen mass among bird species. Data are phylogenetically independent contrasts corrected for bird mass. (From Morand and Poulin 2000)

and Wilson 1999; Schmid-Hempel 2003). Little is known, however, about these costs (Fair et al. 1999; Lochmiller and Deerenberg 2000), other than the negative relationships sometimes observed between investment in reproduction, body condition, and immune responses, or between antigen challenge and growth rate (Fair et al. 1999; Bonneaud et al. 2003; Soler et al. 2003).

According to Tella et al. (2002), the development of a costly immune function should be balanced through the evolutionary optimization of resource allocation between reproduction and mechanisms that promote survival. This trade-off perspective leads to the following four hypotheses based on life-history theory (Stearns 1992): (1) immune function should follow an allometric relationship with body size (Calder 1984); (2) immune function should increase with increasing adult survival rate or, more generally, increasing lifespan; (3) immune function should be directly related to the length of the developmental period if time is required for the maturation of critical components of the immune system (Ricklefs 1992); and (4) immune function should be directly related to the prevalence and/or the diversity of parasites and diseases as a means to control infection (Martin et al. 2001). At this time, these four hypotheses have not been rigorously tested one against another. The last one, however, assigns a role to parasite diversity in the evolution of host immune defenses, and thus in the diversification of host life-history strategies.

Some studies, although not directly testing parasite species richness as a determinant of investment in immunity, have nonetheless found relationships using surrogate measures. For instance, comparative studies suggest that hosts allocate their investment in immune function differently and as a function of their probability of exposure to parasites (Møller et al. 2001; Nunn et al. 2003b). Other factors affecting exposure to parasites can also modulate the evolution of immune function. For example, animals living in large groups or colonies, undergoing extensive migration, or living in the tropics show greater investments in immunity than their counterparts living solitarily or in small groups, not migrating, or living in temperate regions (Møller 1998; Møller and Erritzøe 1998); for various ecological reasons, the former are thought to face greater pressures from parasites than the latter.

One of the major obstacles facing anyone attempting to assess the importance of parasite diversity in shaping the evolution of immune defenses is how exactly to measure these immune responses or the relative investment in immune function. What is needed is a comparative measure of the magnitude of the immune investment made by different host species, a measure that can then be related to the interspecific variation in the diversity of parasites faced by different hosts. At the moment, there are only two ways to achieve this. The first one is to measure the size of organs involved in the immune defense, organs such as the spleen (see John 1994) or the bursa of Fabricius in birds. The second one is to measure a nonspecific lymphocyte proliferation induced by a standard exposure to antigens, such as the injection of PHA (see Smits et al. 1999, 2001). Both measurements are not without biases and pitfalls, but if used carefully they can be instructive.

Following this comparative approach, John (1995) found a positive relationship between the prevalence of nematode parasites and the relative weight of the spleen across bird species. Further, Morand and Poulin (2000) also found a positive relationship between the relative mass of the avian spleen and the nematode species richness (controlled for host sample size and host bird mass) harbored by bird species, supporting the view that the investment in immune organs is dependent on the parasite diversity facing the host. Moreover, Morand and Poulin (2000) confirmed the existence of a trade-off between the investment in reproduction and the investment in immune defenses, as they found a negative relationship between the relative testes mass and the relative spleen mass (Figure 8.3). Other comparative studies add support to the idea that parasites play a role in the evolution of host defenses by showing that bird species that are likely exposed to more parasites, such as migratory species or species that live in the tropics, display relatively larger immune organs (spleen or bursa of Fabricius) than their sedentary or temperate relatives (Møller 1998; Møller and Erritzøe 1998).

Intraspecific variation in spleen sizes sheds a different light on the issue, however. In a comparative study of 368 species of birds, significantly larger spleens were consistently observed in females than in males (controlling for body size) (Møller et al. 1998). These results suggest that males may have lower investment in immune defenses because they allocate relatively more energy to sexual behavior aimed at obtaining mates than females do. This is still consistent with a role for parasites in the evolution of spleen sizes, but other kinds of intraspecific variation are not. For instance, large spleen size can reflect splenomegaly (i.e., accelerated spleen growth) as a response to current levels of parasitism. In such a case, spleen size would not be an indication of evolutionarily optimized patterns of resource allocation to immune defense or other functions. Shutler et al. (1999) found that spleen size covaries with helminth loads in geese, *Chen caerulescens*, with current infection levels explaining much of the intraspecific variation in spleen sizes. Similarly, Brown and Bomberger Brown (2002), investigating spleen volume variation within a bird species as a function of bird colony size and parasite load, found that birds from parasite-free colonies had significantly smaller spleens than birds from parasitized colonies. Their results are more consistent with parasite-induced increases in spleen size as proximate responses to infection than with spleen size being a reliable signal of differential life-history investment in immune defenses. Brown and Bomberger Brown (2002) suggested that comparative analyses across host species should use only spleen sizes from uninfected individuals, data difficult to obtain under natural conditions. Thus, despite the comparative studies suggesting that parasite infection levels or parasite diversity influence the evolution of organs of immune defense (John 1995; Møller 1998; Møller and Erritzøe 1998; Morand and Poulin 2000), conclusive evidence is still missing.

Analyses of cell-mediated immunity based on responses to antigen challenges also suggest a role for parasite diversity in the evolution of immune defenses. The evolution of immunocompetence measured by artificially induced lymphocyte proliferation is the subject of a growing number of studies (see Deerenberg et al. 1997; Christe et al. 2000; Lochmiller and Deerenberg 2000). For example, the phytohemagglutinin (PHA) skin test consists of injecting subcutaneously an antigen and measuring the amount of swelling after a period of 24 hours (Smits et al. 1999, 2001). The response in this case involves macrophages, basophils, heterophils, and B lymphocytes, and it is orchestrated by cytokine secreted by T lymphocytes. Heterophils and macrophages infiltrate the injection site 1 to 2 hours after injection and the final stage of the response consists of a swelling at the injection site caused by dense infiltration of macrophages, lymphocytes, basophils, and heterophils. This test, like others of its kind, provides a general index of cell-mediated immunity when applied in a standard way to several host individuals of the same or different species.

Numerous studies, mostly in birds at either the intra or interspecific level, have investigated the relationships between cell-mediated immunity and host features such as sex (Moreno et al. 2001), diet (Gonzalez et al. 1999), body condition (Alonso-Alvarez and Tella 2001), parental clutch size (Tella et al. 2000), egg-laying date (Hasselquist et al. 2001), reproductive effort (Deerenberg et al. 1997; Moreno et al. 1999; Nordling et al. 1998; Horak et al. 1998), parental workload (i.e., nest provisioning: Raberg et al. 1998; Svensson et al. 1998; Hasselquist et al. 2001), survival rate (Gonzalez et al. 1999; Horak et al. 1999; Sinclair and Lochmiller 2000), and exposure to parasites (Christe et al. 2000).

Comparative analyses looking directly at parasite diversity and cell-mediated immunity are scarce, however. One study related the magnitude of cell-mediated immune responses to nestling mortality in birds, linking immunocompetence with clutch size (Martin et al. 2001). More relevant is a recent study reporting a correlation between the strength of cell-mediated responses and the extent of host sociality among bird species (Møller et al. 2001). This result supports the hypothesis that host species living at high densities should harbor more parasite species than solitary ones (Freeland 1979; Côté and Poulin 1995; see Chapter 3), placing them under greater pressure to fight infection. The same study also reported that bird species breeding in a wide range of habitats had lower immunocompetence than habitat specialists (Møller et al. 2001). This finding does not support the idea that selective pressure due to parasite diversity are important in the evolution of cell-mediated immunity, as hosts living in a range of different habitats should harbor a higher parasite species diversity. Only one study across bird species has clearly shown a positive correlation between parasite species richness and immune defense, measured by the strength of cell-mediated immune responses. Møller and Rozsa (unpublished data) found that, among bird species, the species richness of amblyceran lice was positively correlated with immune responses of nestling birds, but that the richness of ischnoceran lice was not. Amblyceran lice feed on host skin and are believed to be more virulent than ischnoceran species that feed on keratin. In any event, this result and others discussed above suggest, although not convincingly, that the diversity of parasite species exploiting a host population or species may be a key determinant of how the immune system of its members may be evolving.

Parasites and the Evolution of the MHC

Vertebrates have developed an original molecular mechanism for differentiating self and nonself (i.e., parasites and pathogens). This mechanism is based on the recognition of an antigen or its three-dimensional motif, and requires three molecules, the major histocompatibility complex (MHC), the T-cell receptor, and the immunoglobulin molecules (antibodies). The MHC is coded by a large chromosomal region that contains several closely linked and highly polymorphic genes; these play a central role in the vertebrate immune system. Their products, the MHC molecules, control immunological self and nonself recognition by binding foreign proteins and by presenting them to patrolling T lymphocytes (Klein 1986).

Since the MHC is directly involved in the fight against parasites, its diversity may be maintained by pathogen interactions, but it is also often invoked as an inbreeding avoidance mechanism (Brown and Eklund 1994; Apanius et al. 1997; Penn and Potts 1998; Meyer and Thomson 2001). Selection should favor MHC heterozygosity as a way of increasing the effectiveness of the immune system in dealing with a wide variety of pathogens (Nei and Hughes 1991). Pathogens may thus be important agents of balancing selection (Klein 1986). Parasite-driven selection on the MHC can operate at the genetic level in at least two ways: first, via heterozygous advantage (Hughes and Nei 1989), and second, through frequency-dependent selection (Slade and McCallum 1992).

There is empirical support for both selection processes. Heterozygous individuals should detect and respond to a wider range of pathogen-derived antigens than homozygous ones, due to a greater number of different MHC molecules. Two studies have confirmed this hypothesis by showing a heterozygous advantage in HIV infection and hepatitis B viral infection (Thursz et al. 1997; Carrington et al. 1999). The alternative mechanism is negative frequency-dependent selection, where rare MHC alleles have a selective advantage over common alleles. This happens because rapidly evolving parasites track common host genotypes, quickly making them obsolete; only rare MHC genotypes are likely to confer efficient resistance against widespread parasites. This hypothesis has gained support from studies showing correlations between certain MHC genotypes or alleles on the one hand and disease resistance or other fitness traits on the other (Hill et al. 1991; Slade and McCallum 1992; Paterson et al. 1998; Langerfors et al. 2001).

Several alternative mechanisms that could maintain MHC diversity have been suggested: reproductive mechanisms such as negative assortative mating, maternal–fetal interactions to increase reproductive efficiency, and the use of the MHC as olfactory-based markers for kin recognition to avoid inbreeding (Hill et al. 1991; Manning et al. 1992; Brown and Eklund 1994; Hedrick 1994). There is also good evidence for all of these. Mice prefer mates that have MHC genotypes differing from their own (Yamazaki et al. 1976, 1983, 1994; Egid and Brown 1989; Potts et al. 1991, 1994; Jordan and Bruford 1998; Penn and Potts 1998). Odor-based disassortative mating with respect to MHC genotypes has been demonstrated (Wedekind et al. 1996; Wedekind and Füri 1997), and a set of olfactory receptor-like genes has been found to be located in the human MHC (Fan et al. 1995; Ziegler et al. 2000). In addition, recent studies have revealed that neutral forces such as genetic drift, gene flow, and inbreeding might also influence MHC variation in mammalian populations (Boyce et al. 1997; Seddom and Baverstock 1999; Pfau et al. 2001). Balancing selection (Hughes and Yeager 1998; Richman 2000) and reproductive mechanisms (Jordan and Bruford 1998; Penn and Potts 1998), which are thought to act at MHC loci, have been suggested to counteract the effects of genetic drift by maintaining polymorphism within populations. It is therefore clear that although parasites play an important role in the evolution of MHC genotypes, they are only one of many influences on the maintenance of MHC genetic diversity.

Conservation biologists also have recently turned to the study of genetic variability in the MHC. Several hypotheses have been proposed to explain low MHC polymorphism in endangered mammals (Sommer et al. 2002). Some hypotheses invoke the reduced pressure from parasites and pathogens due to low parasite diversity in environments such as in certain marine habitats, as seen for instance in some marine mammals (Trowsdale et al. 1989; Slade 1992). The reduced selective pressure from parasites on MHC polymorphism may also be exacerbated by isolation or solitary life styles, which are generally associated with lower parasite diversity, as in the Swedish moose (*Alces alces*) (McGuire et al. 1985) or the Syrian hamster (*Mesocricetus auratus*) (Ellegren et al. 1996). Hence, the role of parasites in maintaining MHC genetic diversity is of great relevance to conservation biologists worried about disease resistance in endangered populations.

No direct evidence links parasite species diversity and the level of genetic polymorphism of the MHC (but see Wegner et al. 2003), although there is circumstantial evidence. Lower vertebrates (sharks, teleost fish, reptiles) possess a lower number of specific antibodies than birds and mammals: fewer than 500,000 versus 10^7 to 10^9 (Du Pasquier 1982; see Frost 1999). The emergence of the MHC in the lower vertebrates was followed by the dramatic expansion and duplication of MHC genes in birds and mammals (Klein 1991). In parallel to this expansion of the MHC, the selective pressures exerted by parasites have also increased from the lower to the higher vertebrates, as revealed by using parasite species richness as an indicator: helminth species richness is lower in fishes, reptiles, and amphibians than in birds and mammals (see Chapter 4). Combes and Morand (1999) argued that the relationships between the increase of the antibodies repertoire, the expansion of MHC loci, and the selective pressures exerted by parasites strongly support the hypothesis that parasites are the main agents that have driven the evolution of the immune system of vertebrates (Klein 1991).

We are slowly moving toward a general framework for studies of immunoecology—the relationships between immune defense and other functions (Figure 8.4). It is also becoming clear that, although parasite diversity is influencing the evolution of many related aspects of host biology, some of these features of hosts are themselves determining how many parasite species will evolve to exploit the host (see Chapters 3 and 4). The direction of causality is not always clear in these relationships.

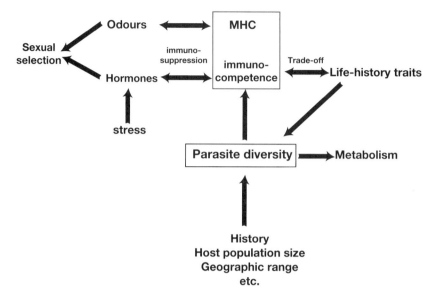

Figure 8.4. Summary of the interrelationships discussed in Chapter 8, providing a general framework of the links between parasite diversity and the evolution of host life history traits through the central immunocompetence component.

Parasites, Host Sex, and Host Sexual Selection

Parasites are now widely believed to have influenced the evolution of sex, as well as the evolution of differences between the sexes. Hamilton and Zuk (1982) proposed that secondary sexual characteristics evolved because they act as indicators of heritable resistance against parasites and diseases. If only resistant males with low parasite loads can produce exaggerated sexual ornamentation, a female can secure good genes (i.e., parasite resistance) for her offspring by choosing highly ornamented mates. Hamilton and Zuk (1982) predicted that among animal species, those subject to greater selective pressures from parasites should have evolved more elaborate secondary sexual traits, such as brighter coloration or more complex songs and displays, than species facing few parasites. There have been many tests of this prediction, in which comparative measures of sexual ornamentation are related to measures of infection levels, usually involving parasite species richness; although the results are not always consistent, the consensus is that parasite diversity has played a role in the evolution of sexually selected traits (see reviews in Clayton 1991; Zuk 1992; Hamilton and Poulin 1997).

In an extension of the ideas of Hamilton and Zuk (1982), Folstad and Karter (1992) proposed that parasites should influence the expression of secondary sexual traits through the interaction between infection, hormones, and the immune system of hosts. Steroid hormones associated with reproduction and stress can mediate relationships between behavior and immune function (Grossman 1985; Maier et al. 1994; Wedekind and Folstad 1994; Beckage 1997; Hillgarth and Wingfield 1997; Wilckens and de Rijk 1997; Maier and Watkins 1999; Barnard and Behnke 2001). A proximate feedback of immunosuppression results from steroid hormones and their precursors having a direct effect on immune cells by decreasing the level of antibodies in the plasma, as seen in fish for instance (Slater et al. 1995; Hou et al. 1999). Sexual characters develop in response to the same hormones that negatively affect the immune system (Folstad and Karter 1992). Males engaged in intense sexual competition are supposed to invest largely in testosterone production, with a coincident immunosuppression side-effect; this forms the basis of the immunohandicap hypothesis (Folstad and Karter 1992; Wedekind and Folstad 1994; Skarstein and Folstad 1996). Such a trade-off between investment in male reproduction and immunocompetence has been reported in many organisms, such as birds, where Poulin and Morand (2000) showed a negative interspecific relationship between the relative size of testes and the relative size of the spleen, an organ involved in immunity (Figure 8.3). Similarly, experimental activation of the immune system is followed by a reduction of the expression of a secondary sexual trait in male blackbirds (Faivre et al. 2003). This brings us back to the idea that any investment in immune defense is made at the expense of investment in other functions, and that the optimal allocation of energy between functions is, to some extent, determined by the number of parasite species exploiting a host species. Parasite diversity selects for levels of host immune defense, and this in turns modulates the evolution of life-history traits and sexually selected characters (Zuk 1996; Møller 1997; Møller et al. 1999; Zuk and Stoehr 2002). In turn, host investment in immunity leads to adaptive changes in parasite life history traits (Sorci et al. 2003).

Finally, parasites may be responsible for the maintenance of sexual reproduction itself in their hosts (Jaenike 1978; Hamilton 1980; Bremermann 1985; Hamilton et al. 1990). Sexual reproduction allows hosts to rearrange their genotypes through outcrossing and recombination and to produce offspring with greater genetic diversity. This hypothesis has become known as the Red Queen hypothesis: sex may allow hosts to keep up with their fast-evolving pathogens in an arms race in which hosts are always lagging behind. Support for this hypothesis has come from a range of studies, on hosts as varied as snails (Lively 1987), fishes (Lively et al. 1990), and lizards (Moritz et al. 1991; Hanley et al. 1995), among others. Parasites may thus be implicated in the evolution of sex, one of the most fundamental biological processes.

Final Words: Parasitism and Host Diversification

This chapter has provided an abridged review of the many ways in which parasites can cause variation in host features among host populations or species. By favoring investment in certain functions, or promoting certain life-history strategies over others, parasites have generated differences among free-living species. But the ultimate question is whether parasites have increased the biodiversity of free-living organisms by promoting higher speciation rates. In other words, would the diversity of free-living organisms be lower had there been fewer parasite species over the past several million years?

The idea that parasites can be associated with host speciation has received attention at various times in the past (Thompson 1987; Poulin and Thomas 1999; Summers et al. 2003). Certain larval helminths induce profound phenotypic changes in their intermediate hosts, sometimes resulting in a clear spatial segregation between parasitized and unparasitized hosts (Poulin and Thomas 1999). Although this can lead to assortative pairing based on infection status, it could not possibly lead to sympatric speciation. Parasite-mediated allopatric speciation, however, is entirely possible when different populations of the same host species differ greatly either in levels of infection by a particular parasite or in the richness of the parasite fauna they harbor. Different selection pressures in these different populations would lead to some divergence in the types of traits favored. Over evolutionary time, this would not be sufficient for allopatric speciation, as the evolution of pre- or post-zygotic isolation mechanisms would be necessary, but it could facilitate it. Bacteria of the *Wolbachia* group are among the best candidates for this sort of parasite-mediated allopatric speciation. These bacteria live in the reproductive tissues of arthropods and cause cytoplasmic incompatibility between eggs and sperm of different individuals. Although this may contribute to the isolation of host strains, there is no conclusive data that *Wolbachia* can induce host speciation (Weeks et al. 2002).

Evidence for the role of parasites in host diversification comes from simple and elegant studies using bacteria, *Pseudomonas fluorescens*, growing in microcosms (Buckling and Rainey 2002; Brockhurst et al. 2004). When propagating in a nutrient-rich culture medium, these bacteria diversify rapidly into different niche specialists distinguished by their colony morphologies. Buckling and Rainey (2002) inoculated some microcosms with a virulent phage—a virus parasite that invades bacterial cells, replicates, and then kills the host cell. They found that bacterial populations evolving without phages tended to contain the same morphotypes at similar frequencies. In contrast, populations evolving with phages were dominated by different morphotypes in different populations. In other words, the diversity of morphotypes among bacterial populations was much higher when phage parasites were present than when they were absent. This example illustrates well how parasites can influence host diversification. Whether other types of parasites, such as helminths or arthropods, can have the same effect, remains to be determined, although this will be impossible to achieve experimentally. Still, it is probably safe to say that the world of free-living organisms that we see with our eyes would be considerably poorer if it were not for parasites.

9

The Study and Value of Parasite Biodiversity

This book has explored the origins, distribution, and consequences of parasite diversity. It should be clear to readers by now that we have only scratched the surface: a lot of questions remain unanswered. When walking through a forest, a biologist sees birds flying among the trees, and explaining why this particular forest is inhabited by a more diverse bird assemblage than another forest is an obvious question to tackle. Most biologists do not see that each bird is in fact a flying community of parasites, both internal and external. In great part because of their hidden or even invisible nature, parasites are commonly overlooked in biodiversity studies. Here we conclude with suggestions for new approaches to the study of parasite diversity and with a discussion of its value as a natural resource to protect.

Parasite Biodiversity: Past, Present, and Future
Research on parasite diversity is still in its infancy, although it appears to be going through a rapid growth phase at the moment. Some indications of this come from recent trends in the scientific literature. For instance, certain landmark papers on the diversity of parasites and its causes, published two decades ago, are cited with increasing rather than decreasing frequency over time. This is the case for the paper by Kuris et al. (1980), the first to provide a robust critical examination of island biogeography theory as a framework for studies of parasite diversity, and for the paper of Price and Clancy (1983), the first comparative analysis of parasite species richness in fish, the group of hosts that has received the most attention (see Figure 9.1). The citation rate of these papers has increased since the 1980s. Similarly,

the actual number of studies on parasite diversity has increased at about the same time. Looking through the studies reviewed in the comprehensive literature survey presented in this book, and focusing only on comparative studies (across host populations or species, across host habitats, or among parasite taxa) that tested for the effects of putative factors involved in the diversification of parasite faunas or parasite clades, we find that such studies have become much more numerous in the 1990s (Figure 9.1). Paralleling these developments, the rise of molecular approaches to parasite systematics has allowed further insights into parasite biodiversity. Research on parasite diversity has therefore just reached its cruising speed, and the immediate future should provide answers to the many questions still left unresolved. For this to happen, however, two things are necessary: we must put more emphasis on facets of parasite biodiversity other than just species richness, and we must make a concerted effort to create comprehensive, easily accessible databases that will bring together existing and new information on parasite diversity.

First, we need to adopt new approaches to measuring parasite biodiversity. This book used species richness as a measure of biodiversity. We were forced to do this, knowing very well that biodiversity can be quantified in a number of ways, because almost all empirical studies to date have used this measure (Purvis and Hector 2000). It is likely, however, that the use of more sophisticated indices of diversity would reveal patterns that an analysis restricted to species richness would have missed. Consider the two examples in Figure 9.2. Both host species harbor seven parasite species and thus have parasite assemblages of identical species richness. From an ecological perspective, however, we notice that the first host species harbors equal numbers of individuals of each of its seven species, whereas the second host species has a much less even distribution of parasite numbers among its species, with one parasite showing a clear numerical dominance over the others (Figure 9.2). A wide range of indices exist that quantify species diversity by taking into account the evenness or equitability with which individuals are distributed among species (Magurran 1988). For instance, with the Shannon diversity index, the parasite assemblage in the first host scores 1.945, close to the maximum index value for an assemblage with seven species ($\ln(7) = 1.946$), whereas the second species has a lower value of 1.148. This index indicates that the first host species harbors a more diverse parasite assemblage than the second because the latter's assemblage is

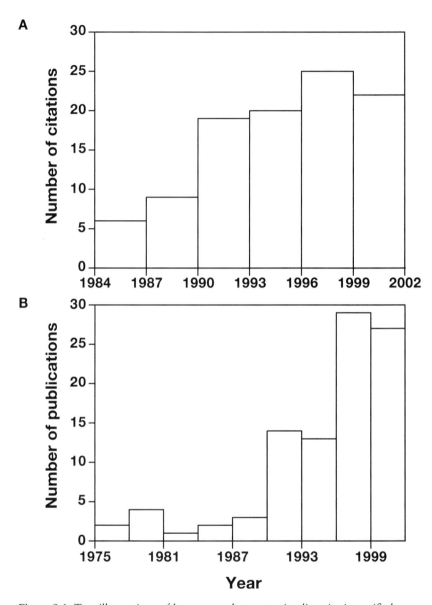

Figure 9.1. Two illustrations of how research on parasite diversity intensified around 1990. (A) Number of times the pioneering papers of Kuris et al. (1980) and Price and Clancy (1983) were cited in the scientific literature in three-year intervals, according to ISI's Science Citation Index. The citations for both papers are combined in the figure. (B) Number of scientific articles on parasite diversity (see text for details) published during three-year intervals since 1975.

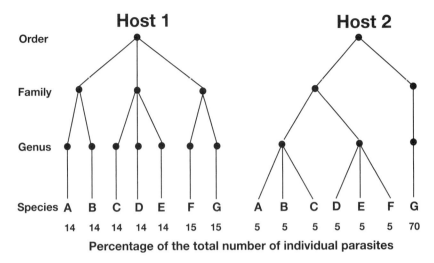

Figure 9.2. Hypothetical parasite assemblages in two host species, each comprising seven parasite species A to G. The percentages of the total number of individual parasites in the assemblage represented by each species is shown below; the taxonomic tree of the parasite species, indicating their relationships, is shown above.

dominated by a single species. Some investigators have used diversity indices of one kind or another in studies of parasite diversity (e.g., Watve and Sukumar 1995), but available studies are still few.

We can also adopt an evolutionary perspective when looking at the examples in Figure 9.2 and focus on the phylogenetic origins of the parasite species in the assemblage. There are no congeneric species in the parasite assemblage of the first host, whereas two genera contribute most of the species in the parasite assemblage of the second host. From a phylogenetic or taxonomic perspective, the first assemblage appears more diverse than the second, because generally we have to go further up the taxonomic tree to find a node common to any two parasite species. Recent indices allow the taxonomic distinctness of species assemblages to be quantified rapidly and turned into a simple index (Clarke and Warwick 1998; Warwick and Clarke 2001). For instance, an index of taxonomic distinctness can simply be the mean number of steps up the taxonomic hierarchy depicted in Figure 9.2 that must be taken to reach a taxon common to two parasite species, computed across all possible species pairs. With each step between nodes having a value of 1, we get index values of 2.64 and 1.91 for the first and second host, respectively. The higher index value reflects the greater taxonomic distance, on average, between parasite species of the first host than those of the second host. The taxonomic tree can be replaced by a proper phylogeny in which branch lengths are used instead of standard step lengths. In an analysis of helminth diversity in mammalian host species, this measure of taxonomic distinctness proved to be a more reliable and repeatable index of parasite diversity than species richness: its variance was much lower among populations of the same host species than among host species, whereas this was not true of species richness (Poulin and Mouillot 2004). The point here is that it is possible to view parasite diversity from different angles, and that similarity in parasite species richness can hide profound differences in either ecological or phylogenetic diversity. The use of indices of taxonomic distinctness or related evolutionary-biased measures of diversity is particularly promising in the context of the ideas presented in this book. To our knowledge, they have only been applied to parasite assemblages in three studies (Poulin and Mouillot 2003, 2004; Luque et al. 2004), at least as this book went to press. We encourage future investigators of parasite biodiversity to embrace these or other (e.g., Petchey and Gaston 2002) recently-developed measures of biodiversity, in the hope that they will tell us more than species richness has.

The second way to promote and facilitate studies of parasite biodiversity is to create public databases bringing together all existing data on parasite diversity in the scientific literature. The Human Genome Project is an example of the power of such an approach. On a smaller scale, the deposition of gene sequences in GenBank is providing phylogeneticists with large, easily searched datasets from which to reconstruct the evolutionary histories of particular groups. In the context of parasite biodiversity, hundreds of studies on the metazoan parasites of vertebrates have been published in many languages and in numerous journals. Because of ethical and conservation restrictions now in place in many countries, obtaining new data is often impossible, so we must make the most of the existing data. A database, accessible via the Internet, in which each sample of vertebrate hosts examined for parasites would be entered, would provide a huge step forward for studies of parasite biodiversity. For each entry, the basic information required would be simple: host species name, numbers examined, site of collection, parasite species found, prevalence and mean intensities of infection, and so on. Searching through the database by host taxa or geographic location would allow many of the hypotheses presented earlier in this book, as well as many new ones, to be tested using a greatly enlarged dataset. The limited data that single researchers working on their own can hope to accumulate has proven a major obstacle in the past; building a collective database is the only solution. A similar rationale for the electronic compilation of the Earth's biodiversity has been presented in other contexts (e.g., Wilson 2003); advancing our knowledge of parasite biodiversity would be only one of many benefits.

Why Bother About Parasite Diversity?

We should soon know more about parasite biodiversity and obtain answers to all the questions posed throughout this book. But why should we care about preserving parasite biodiversity in the first place? Is it really necessary for our planet's life-support systems? Given how difficult it is to excite people about the conservation of plants and animals in general (except for large mammals and showy birds—the charismatic megafauna), generating support for the study and maintenance of parasite biodiversity must be a hopeless task. There are, however, at least two reasons for caring.

First, we can adopt an ethical position. We are the self-appointed guardians of the planet, and the responsibility of this stewardship means that we have a moral duty to look after it for future generations. This involves looking after all other living organisms, be they large or small, free-living or parasitic. We find this a most compelling argument on its own, but we realize that most people, including many biologists, will not be won over so easily and will be convinced only by the demonstration that there are direct benefits to be gained from a diverse parasite fauna.

Therefore the second reason we offer for studying and preserving parasite biodiversity is a utilitarian one. On a broad scale, there is now overwhelming evidence that parasites affect ecosystem processes by regulating host population abundance, modulating interspecific interactions, and affecting the structure and stability of natural communities (Minchella and Scott 1991; Huxham et al. 1995; Combes 1996; Morand and Gonzales 1997; Hudson and Greenman 1998; Hudson et al. 2002; Mouritsen and Poulin 2002). Clearly, they are an integral part of the biosphere and play a role in its functioning. Just as parasites have promoted the diversity of living organisms over evolutionary time (Chapter 8), they can help sustain diverse ecosystems now, maintaining the Earth as a place where life can flourish.

At a more applied level, the planet's biodiversity in general has recently been recognized as a vast source of products to be exploited commercially (ten Kate and Laird 1999), and there may also be potential industrial, agricultural, or medical applications to be derived from knowledge of parasite biology and diversity. We still have few concrete examples, but the possibilities are exciting. For instance, several parasites are already proving efficient as biological control agents in managing insect pests in agriculture, where alternatives to chemical agents are urgently needed. Benign strains of helminths parasitic on humans could be used as drug delivery systems, an application that would take advantage of their site specificity and relatively long life. We are just starting to work our way through the parasite's own pharmacopeia. Hookworms, ticks, and leeches all produce molecules that have potential uses as anticoagulants and blood thinners for surgery. Fungi parasitic on insects produce cyclosporin and other substances with antibiotic or anticancer properties, or they may be used as immunosuppressive agents during organ transplants (Jong et al. 1991; Wang and Shiao 2000). Nematodes parasitic on insects, or their bacterial symbionts to be more precise, are also a source of novel antibiotics (Webster et al. 1998; Isaacson and Webster 2002). Just as plants are proving a bountiful source of drugs, parasites and other invertebrates also possess a rich biochemistry, mostly untapped, awaiting investigation and application.

Our hope with this book was that it would not only summarize the state of knowledge on parasite biodiversity, but would stimulate others to take up the challenge and tackle the many issues still unresolved. Studies of parasites can shed more light on biodiversity theories than those on free-living organisms because parasite assemblages occur in well-defined habitats and within a spatial structure (host individual, population, community) amenable to more rigorous analysis. They represent the unseen biodiversity *within* the biodiversity that we see with our eyes, a whole other level of diversity that contributes to shaping the one we see. Understanding what drives this unseen diversity is a fundamental step toward understanding our living world.

References

Acevedo, D.H. and Currie, D.J. 2003. Does climate determine broad-scale patterns of species richness? A test of the causal link by natural experiment. *Global Ecology and Biogeography* 12: 461–473.

Adlard, R.D., Barker, S.C., Blair, D. and Cribb, T.H. 1993. Comparison of the second internal transcribed spacer (ribosomal RNA) from populations and species of Fasciolidae (Digenea). *International Journal for Parasitology* 23: 423–425.

Agapow, P.-M. and Isaac, N.J.B. 2002. MacroCAIC: revealing correlates of species richness by comparative analysis. *Diversity and Distributions* 8: 41–43.

Aho, J.M. 1990. Helminth communities of amphibians and reptiles: comparative approaches to understanding patterns and processes. In: *Parasite Communities: Patterns and Processes* (Esch, G.W., Bush, A.O. and Aho, J.M., editors). Chapman and Hall, London, pp. 157–195.

Aho, J.M. and Bush, A.O. 1993. Community richness in parasites of some freshwater fishes from North America. In: *Species Diversity in Ecological Communities: Historical and Geographical Perspectives* (Ricklefs, R.E. and Schluter, D., editors). University of Chicago Press, Chicago, pp. 185–193.

Allen, A.P., Brown, J.H. and Gillooly, J.F. 2002. Global biodiversity, biochemical kinetics, and the energetic-equivalence rule. *Science* 297: 1545–1548.

Alonso-Alvarez, C. and Tella, J.L. 2001. Effects of experimental food restriction and body-mass changes on the avian T-cell-mediated immune response. *Canadian Journal of Zoology* 79: 101–105.

Alroy, J. 2002. How many named species are valid? *Proceedings of the National Academy of Sciences U.S.A.* 99: 3706–3711.

Altizer, S., Nunn, C.L., Thrall, P.H., Gittleman, J.L., Antonovics, J., Cunningham, A.A., Dobson, A.P., Ezenwa, V., Jones, K.E., Pedersen, A.B., Poss, M. and Pulliam, J.R.C. 2003. Social organization and parasite risk in mammals: integrating theory and empirical studies. *Annual Review of Ecology, Evolution, and Systematics* 34: 517–547.

Amin, O.M. 1975. Variability in *Acanthocephalus parksidei* Amin, 1974 (Acanthocephala: Echinorhynchidae). *Journal of Parasitology* 61: 307–317.

Amin, O.M. 1985. Classification. In: *Biology of the Acanthocephala* (Crompton, D.W.T. and Nickol, B.B., editors). Cambridge University Press, Cambridge, pp. 27–72.

Amin, O.M. 1987. Key to the families and subfamilies of Acanthocephala, with the erection of a new class (Polyacanthocephala) and a new order (Polyacanthorhynchida). *Journal of Parasitology* 73: 1216–1219.

Andersen, K.I. and Valtonen, E.T. 1990. On the infracommunity structure of adult cestodes in freshwater fishes. *Parasitology* 101: 257–264.

Anderson, R.C. 2000. *Nematode Parasites of Vertebrates: Their Development and Transmission,* second edition. CAB International, Wallingford, UK.

Anderson, R.M. and May, R.M. 1978. Regulation and stability of host-parasite population interactions. I. Regulatory processes. *Journal of Animal Ecology* 47: 219–247.

Anderson, R.M. and May, R.M. 1985. Helminth infections of humans: mathematical models, population dynamics, and control. *Advances in Parasitology* 24: 1–101.

Anderson, R.M. and May, R.M. 1991. *Infectious Diseases of Humans: Dynamics and Control.* Oxford University Press, Oxford.

Apanius, V., Penn, D., Slev, P.R., Ruff L.R. and Potts, W.K. 1997. The nature of selection on the major histocompatibility complex. *Critical Reviews in Immunology* 17: 179–224.

Arneberg, P. 2001. An ecological law and its macroecological consequences as revealed by studies of relationships between host densities and parasite prevalence. *Ecography* 24: 352–358.

Arneberg, P. 2002. Host population density and body mass as determinants of species richness in parasite communities: comparative analyses of directly transmitted nematodes of mammals. *Ecography* 25: 88–94.

Arneberg, P., Skorping, A., Grenfell, B.T. and Read, A.F. 1998. Host densities as determinants of abundance in parasite communities. *Proceedings of the Royal Society of London* B 265: 1283–1289.

Ashford, R.W. and Crewe, W. 2003. *The Parasites of* Homo sapiens, second edition. Taylor and Francis, London.

Ba, C.T., Wang, X.Q., Renaud, F., Euzet, L., Marchand, B. and de Meeûs, T. 1993. Diversity and specificity in cestodes of the genus *Moniezia:* genetic evidence. *International Journal for Parasitology* 23: 853–857.

Baltanás, A. 1992. On the use of some methods for the estimation of species richness. *Oikos* 65: 484–492.

Barker, D.E., Marcogliese, D.J. and Cone, D.K. 1996. On the distribution and abundance of eel parasites in Nova Scotia: local versus regional patterns. *Journal of Parasitology* 82: 697–701.

Barker, S.C. 1991. Evolution of host-parasite associations among species of lice and rock-wallabies: coevolution? *International Journal for Parasitology* 21: 497–501.

Barker, S.C. 1994. Phylogeny and classification, origins, and evolution of host associations of lice. *International Journal for Parasitology* 24: 1285–1291.

Barnard, C.J. and Behnke, J.M. 2001. From psychoneuroimmunology to ecological immunology: life history strategies and immunity trade-offs. In: *Psychoneuroimmunoecology* (Ader, R., Felten, D. and Cohen, N., editors). Academic Press, San Diego, pp. 34–47.

Barnes, R.S.K. 1998. *The Diversity of Living Organisms*. Blackwell Science, Oxford.

Barraclough, T.G., Vogler, A. and Harvey, P.H. 1998. Revealing the factors that promote speciation. *Philosophical Transactions of the Royal Society of London* B 353: 241–249.

Bartoli, P., Bray, R.A. and Gibson, D.I. 1989. The Opecoelidae (Digenea) of sparid fishes of the western Mediterranean. III. *Macvicaria* Gibson and Bray, 1982. *Systematic Parasitology* 13: 167–192.

Beckage, N.E. 1997. *Parasites and Pathogens: Effects on Host Hormones and Behavior*. Chapman and Hall, New York.

Bell, A.S., Sommerville, C. and Gibson, D.I. 2002. Multivariate analyses of morphometrical features from *Apatemon gracilis* (Rudolphi, 1819) Szidat, 1928 and *A. annuligerum* (v. Nordmann, 1832) (Digenea: Strigeidae) metacercariae. *Systematic Parasitology* 51: 121–133.

Bell, G. 2001. Neutral macroecology. *Science* 293: 2413–2418.

Bell, G. and Burt, A. 1991. The comparative biology of parasite species diversity: internal helminths of freshwater fish. *Journal of Animal Ecology* 60: 1047–1064.

Bergeron, M., Marcogliese, D.J. and Magnan, P. 1997. The parasite fauna of brook trout, *Salvelinus fontinalis* (Mitchill), in relation to lake morphometrics and the introduction of creek chub, *Semotilus atromaculatus* (Mitchill). *Ecoscience* 4: 427–436.

Berlocher, S.H. and Feder, J.L. 2002. Sympatric speciation in phytophagous insects: moving beyond controversy? *Annual Review in Entomology* 47: 773–815.

Berra, T.M. 1997. Some 20th century fish discoveries. *Environmental Biology of Fishes* 50: 1–12.

Beveridge, I., Chilton, N.B. and Spratt, D.M. 2002. The occurrence of species flocks in the nematode genus *Cloacina* (Strongyloidea: Cloacininae), parasitic in the stomachs of kangaroos and wallabies. *Australian Journal of Zoology* 50: 597–620.

Beveridge, I. and Jones, M.K. 2002. Diversity and biogeographical relationships of the Australian cestode fauna. *International Journal for Parasitology* 32: 343–351.

Beveridge, I. and Spratt, D.M. 1996. The helminth fauna of Australasian marsupials: origins and evolutionary biology. *Advances in Parasitology* 37: 135–254.

Blackburn, T.M. and Gaston, K.J. 1994a. Animal body size distributions: patterns, mechanisms and implications. *Trends in Ecology and Evolution* 9: 471–474.

Blackburn, T.M. and Gaston, K.J. 1994b. Animal body size distributions change as more species are described. *Proceedings of the Royal Society of London* B 257: 293–297.

Blackburn, T.M. and Gaston, K.J. 1995. What determines the probability of discovering a species? A study of South American oscine passerine birds. *Journal of Biogeography* 22: 7–14.

Blankespoor, H.D. 1974. Host-induced variation in *Plagiorchis noblei* Park, 1936 (Plagiorchiidae: Trematoda). *American Midland Naturalist* 92: 415–433.

Blaxter, M.L., De Ley, P., Garey, J.R., Liu, L.X., Scheldeman, P., Vierstraete, A., Vanfleteren, J.R., Mackey, L.Y., Dorris, M., Frisse, L.M., Vida, J.T. and Thomas, W.K. 1998. A molecular evolutionary framework for the phylum Nematoda. *Nature* 392: 71–75.

Blouin, M.S. 2002. Molecular prospecting for cryptic species of nematodes: mitochondrial DNA versus internal transcribed spacer. *International Journal for Parasitology* 32: 527–531.

Bonneaud, C., Mazuc, J., Gonzalez, G., Haussy, C., Chastel, O., Faivre, B. and Sorci, G. 2003. Assessing the cost of mounting an immune response. *American Naturalist* 161: 367–379.

Boyce, W.M., Hedrick, P.W., Muggli-Cockett, N.E., Kalinowski, S., Penedo, M.C.T. and Ramey, R.R. II. 1997. Genetic variation of major histocompatibility complex and microsatellite loci: a comparison in bighorn sheep. *Genetics* 145: 421–433.

Brandle, M. and Brandl, R. 2001. Species richness of insects and mites on trees: expanding Southwood. *Journal of Animal Ecology* 70: 491–504.

Bray, R.A. 1986. Patterns in the evolution of marine helminths. In: *Parasitology—Quo Vadit?* (Howell, N.J., editor). Proceedings of the Sixth International Congress of Parasitology. Australian Academy of Science, Canberra, pp. 337–344.

Bray, R.A. and des Clers, S.A. 1992. Multivariate analyses of metrical features in *Lepidapedon elongatum* (Lebour, 1908) species-complex (Digenea: Lepocreadiidae) in deep and shallow water gadiform fish of NE Atlantic. *Systematic Parasitology* 21: 223–232.

Bray, R.A., Littlewood, D.T.J., Herniou, E.A., Williams, B. and Henderson, R.E. 1999. Digenean parasites of deep-sea teleosts: a review and case studies of intrageneric phylogenies. *Parasitology* 119: S125–S144.

Bremermann, H.J. 1985. The adaptive significance of sexuality. *Experientia* 41: 1245–1255.

Brockhurst, M.A., Rainey, P.B. and Buckling, A. 2004. The effect of spatial heterogeneity and parasites on the evolution of host diversity. *Proceedings of the Royal Society of London* B 271: 107–111.

Brooks, D.R. 1992. Origins, diversification, and historical structure of the helminth fauna inhabiting neotropical freshwater stingrays (Potamotrygonidae). *Journal of Parasitology* 78: 588–595.

Brooks, D.R. and Hoberg, E.P. 2000. Triage for the biosphere: the need and rationale for taxonomic inventories and phylogenetic studies of parasites. *Comparative Parasitology* 67: 1–25.

Brooks, D.R. and Hoberg, E.P. 2001. Parasite systematics in the 21st century: opportunities and obstacles. *Trends in Parasitology* 17: 273–275.

Brooks, D.R. and McLennan, D.A. 1991. *Phylogeny, Ecology, and Behavior: A Research Program in Comparative Biology.* University of Chicago Press, Chicago.

Brooks, D.R. and McLennan, D.A. 1993a. Comparative study of adaptive radiations with an example using parasitic flatworms (Platyhelminthes: Cercomeria). *American Naturalist* 142: 755–778.

Brooks, D.R. and McLennan, D.A. 1993b. *Parascript: Parasites and the Language of Evolution.* Smithsonian Institution Press, Washington.

Brown, C.R. and Bomberger Brown, M. 1986. Ectoparasitism as a cost of coloniality in cliff swallows (*Hirundo pyrrhonota*). *Ecology* 67: 1206–1218.

Brown, C.R. and Bomberger Brown, M. 2002. Spleen volume varies with colony size and parasite load in a colonial bird. *Proceedings of the Royal Society of London* B 269: 1367–1373.

Brown, J.H. 1995. *Macroecology.* University of Chicago Press, Chicago.

Brown, J.H. and Lomolino, M.V. 1998. *Biogeography,* second edition. Sinauer Press, Sunderland, MA.

Brown, J.H., Marquet, P.A. and Taper, M.L. 1993. Evolution of body size: consequences of an energetic definition of fitness. *American Naturalist* 142: 573–584.

Brown, J.L. and Eklund, A. 1994. Kin recognition and the major histocompatibility complex: an integrative overview. *American Naturalist* 143: 435–461.

Brusca, R.C. and Wilson, G.D.F. 1991. A phylogenetic analysis of the Isopoda with some classificatory recommendations. *Memoirs of the Queensland Museum* 31: 143–204.

Buckling, A. and Rainey, P.B. 2002. The role of parasites in sympatric and allopatric host diversification. *Nature* 420: 496–499.

Bucknell, D., Hoste, H., Gasser, R.B. and Beveridge, I. 1996. The structure of the community of strongyloid nematodes of domestic equids. *Journal of Helminthology* 70: 185–192.

Bunge, J. and Fitzpatrick, M. 1993. Estimating the number of species: a review. *Journal of the American Statistical Association* 88: 364–373.

Burnham, K.P. and Overton, W.S. 1979. Robust estimation of population size when capture probabilities vary among animals. *Ecology* 60: 927–936.

Bush, A.O. 1990. Helminth communities in avian hosts: determinants of pattern. In: *Parasite Communities: Patterns and Processes* (Esch, G.W., Bush, A.O. and Aho, J.M., editors). Chapman and Hall, London, pp. 197–232.

Bush, A.O., Aho, J.M. and Kennedy, C.R. 1990. Ecological versus phylogenetic determinants of helminth parasite community richness. *Evolutionary Ecology* 4: 1–20.

Bush, A.O., Fernández, J.C., Esch, G.W. and Seed, J.R. 2001. *Parasitism: The Diversity and Ecology of Animal Parasites.* Cambridge University Press, Cambridge.

Bush, A.O. and Holmes, J.C. 1986a. Intestinal helminths of lesser scaup ducks: patterns of association. *Canadian Journal of Zoology* 64: 132–141.

Bush, A.O. and Holmes, J.C. 1986b. Intestinal helminths of lesser scaup ducks: an interactive community. *Canadian Journal of Zoology* 64: 142–152.

Bush, A.O. and Kennedy, C.R. 1994. Host fragmentation and helminth parasites: hedging your bets against extinction. *International Journal for Parasitology* 24: 1333–1343.

Bush, A.O., Lafferty, K.D., Lotz, J.M. and Shostak, A.W. 1997. Parasitology meets ecology on its own terms: Margolis et al. revisited. *Journal of Parasitology* 83: 575–583.

Bush, G.L. 1969. Sympatric host race formation and speciation in frugivorous flies of the genus *Rhagoletis* (Diptera: Tephritidae). *Evolution* 23: 237–251.

Bush, G.L. 1994. Sympatric speciation in animals: new wine in old bottles. *Trends in Ecology and Evolution* 9: 285–288.

Cabaret, J. and Schmidt, E. 2001. Species diversity of nematode communities in the digestive tract of domestic ruminants: multivariate versus univariate estimations. *Parasitology Research* 87: 311–316.

Cabrero-Sañudo, F.J. and Lobo, J.M. 2003. Estimating the number of species not yet described and their characteristics: the case of Western Palaearctic dung beetle species (Coleoptera, Scarabaeoidea). *Biodiversity and Conservation* 12: 147–166.

Calder, W.A.I. 1984. *Size, Function, and Life History.* Harvard University Press, Cambridge, MA.

Campbell, R.A. 1990. Deep water parasites. *Annales de Parasitologie Humaine et Comparée* 65 (Suppl. 1): 65–68.

Campbell, R.A., Haedrich, R.L. and Munroe, T.A. 1980. Parasitism and ecological relationships among deep-sea benthic fishes. *Marine Biology* 57: 301–313.

Canning, E.U. and Okamura, B. 2004. Biodiversity and evolution of the Myxozoa. *Advances in Parasitology* 56: 43–131.

Cardillo, M. 1999. Latitude and rates of diversification in birds and butterflies. *Proceedings of the Royal Society of London* B 266: 1221–1225.

Carlton, J.T. and Geller, J.B. 1993. Ecological roulette: the global transport of nonindigenous marine organisms. *Science* 266: 78–82.

Carney, J.P. and Dick, T.A. 2000. Parasite biogeography: a review of the origins and ideas with specific examples from Holarctic fishes. *Vie et Milieu* 50: 221–243.

Caro, A., Combes, C. and Euzet, L. 1997. What makes a fish a suitable host for Monogenea in the Mediterranean? *Journal of Helminthology* 71: 203–210.

Carrington, M., Nelson, G.W., Martin, M.P., Kinssner, T., Vlahov, D., Goedert, J.J., Kaslow, R., Buchabinder, S., Hoots, K. and O'Brien, S.J. 1999. HLA and HIV-1 heterozygote advantage and B*35–Cw*04 disadvantage. *Science* 283: 1748–1752.

Chao, A. 1987. Estimating the population size for capture-recapture data with unequal catchability. *Biometrics* 43: 783–791.

Charnov, E.L. 1993. *Life History Invariants: Some Explorations of Symmetry in Evolutionary Ecology.* Oxford University Press, Oxford.

Chilton, N.B., Bao-Zhen, Q., Bøgh, H.O. and Nansen, P. 1999. An electrophoretic comparison of *Schistosoma japonicum* (Trematoda) from different provinces in the People's Republic of China suggests the existence of cryptic species. *Parasitology* 119: 375–383.

Chilton, N.B., Beveridge, I. and Andrews, R.H. 1992. Detection by allozyme electrophoresis of cryptic species of *Hypodontus macropi* (Nematoda: Strongyloidea) from macropodid marsupials. *International Journal for Parasitology* 22: 271–279.

Choudhury, A. and Dick, T.A. 2000. Richness and diversity of helminth communities in tropical freshwater fishes: empirical evidence. *Journal of Biogeography* 27: 935–956.

Choudhury, A. and Dick, T.A. 2001. Sturgeons (Chondrostei: Acipenseridae) and their metazoan parasites: patterns and processes in historical biogeography. *Journal of Biogeography* 28: 1411–1439.

Christe, P., Arlettaz, R. and Vogel, P. 2000. Variation in intensity of a parasitic mite (*Spinturnix myoti*) in relation to the reproductive cycle and immunocompetence of its bat host (*Myotis myotis*). *Ecology Letters* 3: 207–212.

Clarke, K.R. and Warwick, R.M. 1998. A taxonomic distinctness index and its statistical properties. *Journal of Applied Ecology* 35: 523–531.

Clayton, D.H. 1991. The influence of parasites on host sexual selection. *Parasitology Today* 7: 329–334.

Clayton, D.H., Price, R.D. and Page, R.D.M. 1996. Revision of *Dennyus* (*Collodennyus*) lice (Phthiraptera: Menoponidae) from swiftlets, with descriptions of new taxa and a comparison of host-parasite relationships. *Systematic Entomology* 21: 179–204.

Clayton, D.H. and Walther, B.A. 2001. Influence of host ecology and morphology on the diversity of Neotropical bird lice. *Oikos* 94: 455–467.

Cleaveland, S., Laurenson, M.K. and Taylor, L.H. 2001. Diseases of humans and their domestic mammals: pathogen characteristics, host range and the risk of emergence. *Philosophical Transactions of the Royal Society of London* B 356: 991–999.

Coe, C.L. 2002. Neuroendocrines and behavioral influences on the immune system. In: *Behavioral Endocrinology* (Becker, J.B., Breedlove, S.M., Crews, D. and McCarthy, M.M., editors). Massachusetts Institute of Technology Press, Cambridge, MA, pp. 373–407.

Coltman, D.W., Pilkington, J.G., Smith, J.A. and Pemberton, J.M. 1999. Parasite-mediated selection against inbred Soay sheep in a free-living, island population. *Evolution* 53: 1259–1267.

Colwell, R.K. and Coddington, J.A. 1994. Estimating terrestrial biodiversity through extrapolation. *Philosophical Transactions of the Royal Society of London* B 345: 101–118.

Combes, C. 1990. Where do human schistosomes come from? An evolutionary approach. *Trends in Ecology and Evolution* 5: 334–337.

Combes, C. 1995. *Interactions Durables: Ecologie et Evolution du Parasitisme.* Masson, Paris.

Combes, C. 1996. Parasites, biodiversity and ecosystem stability. *Biodiversity and Conservation* 5: 953–962.

Combes, C. 2001. *Parasitism: The Ecology and Evolution of Intimate Interactions.* University of Chicago Press, Chicago.

Combes, C. and Morand, S. 1999. Do parasites live in extreme environments? Constructing hostile niches and living in them. *Parasitology* 119: S107–S110.

Connor, E.F. and McCoy, E.D. 1979. The statistics and biology of the species-area relationship. *American Naturalist* 113: 791–833.

Cornell, H.V. and Lawton, J.H. 1992. Species interactions, local and regional processes, and limits to the richness of ecological communities: a theoretical perspective. *Journal of Animal Ecology* 61: 1–12.

Cort, W.W., McMullen, D.B. and Brackett, S. 1937. Ecological studies on the cercariae in *Stagnicola emarginata angulata* (Sowerby) in the Douglas Lake region, Michigan. *Journal of Parasitology* 23: 504–532.

Côté, I.M. and Poulin, R. 1995. Parasitism and group size in social animals: a meta-analysis. *Behavioral Ecology* 6: 159–165.

Cox, C.B. and Moore, P.D. 1993. *Biogeography: An Ecological and Evolutionary Approach,* fifth edition. Blackwell Science, Oxford.

Coyne, J.A. and Price, T.D. 2000. Little evidence for sympatric speciation in island birds. *Evolution* 54: 2166–2171.

Cressey, R. and Boyle, H. 1978. A new genus and species of parasitic copepod (Pandaridae) from a unique new shark. *Pacific Science* 32: 25–30.

Cribb, T.H. 1998. The diversity of the Digenea of Australian animals. *International Journal for Parasitology* 28: 899–911.

Cribb, T.H. 2004. Living on the edge: parasite taxonomy in Australia. *International Journal for Parasitology* 34: 117–123.

Cribb, T.H., Bray, R.A., Wright, T. and Pichelin, S. 2002a. The trematodes of groupers (Serranidae: Epinephelinae): knowledge, nature and evolution. *Parasitology* 124: S23–S42.

Cribb, T.H., Chisholm, L.A. and Bray, R.A. 2002b. Diversity in the Monogenea and Digenea: does lifestyle matter? *International Journal for Parasitology* 32: 321–328.

Crompton, D.W.T. 1999. How much human helminthiasis is there in the world? *Journal of Parasitology* 85: 397–403.

Cumming, G.S. 2000. Using habitat models to map diversity: pan-African species richness of ticks (Acari: Ixodida). *Journal of Biogeography* 27: 425–440.

Curran, S. and Caira, J.N. 1995. Attachment site specificity and the tapeworm assemblage in the spiral intestine of the blue shark (*Prionace glauca*). *Journal of Parasitology* 81: 149–157.

Currie, D.J. 1991. Energy and large-scale patterns of animal- and plant-species richness. *American Naturalist* 137: 27–49.

Curtis, M.A. and Rau, M.E. 1980. The geographical distribution of diplostomiasis (Trematoda: Strigeidae) in fishes from northern Quebec, Canada, in relation to the calcium ion concentrations of lakes. *Canadian Journal of Zoology* 58: 1390–1394.

Dabert, J., Dabert, M. and Mironov, S.V. 2001. Phylogeny of feather mite subfamily Avenzoariinae (Acari: Analgoidea: Avenzoariidae) inferred from combined analyses of molecular and morphological data. *Molecular Phylogenetics and Evolution* 20: 124–135.

Dallas, J.F., Irvine, R.J. and Halvorsen, O. 2001. DNA evidence that *Marshallagia marshalli* Ransom, 1907 and *M. occidentalis* Ransom, 1907 (Nematoda: Ostertagiinae) from Svalbard reindeer are conspecific. *Systematic Parasitology* 50: 101–103.

Damuth, J. 1987. Interspecific allometry of population density in mammals and other animals: the independence of body mass and population energy-use. *Biological Journal of the Linnean Society* 31: 193–246.

Dash, K.M. 1981. Interaction between *Oesophagostomum columbianum* and *Oesophagostomum venulosum* in sheep. *International Journal for Parasitology* 11: 201–207.

Davis, G.M. and Fuller, S.L.H. 1981. Genetic relationships among recent Unionacea of North America. *Malacologia* 20: 217–253.

de Buron, I. and Morand, S. 2004. Deep-sea hydrothermal vent parasites: why do we not find more? *Parasitology* 128: 1–6.

Deerenberg, C., Apanius, V., Daan, S. and Bos, N. 1997. Reproductive effort decreases antibody responsiveness. *Proceedings of the Royal Society of London* B 264: 1021–1029.

Delahay, R.J., Speakman, J.R. and Moss, R. 1995. The energetic consequences of parasitism: effects of a developing infection of *Trichostrongylus tenuis* (Nematoda) on red grouse (*Lagopus lagopus scoticus*) energy-balance, body-weight and condition. *Parasitology* 110: 473–482.

de Meeûs, T., Durand, P. and Renaud, F. 2003. Species concepts: what for? *Trends in Parasitology* 19: 425–427.

de Meeûs, T., Hochberg, M.E. and Renaud, F. 1995. Maintenance of two genetic entities by habitat selection. *Evolutionary Ecology* 9: 131–138.

de Meeûs, T., Michalakis, Y. and Renaud, F. 1998. Santa Rosalia revisited: or why are there so many kinds of parasites in 'The garden of Earthly Delights'? *Parasitology Today* 14: 10–13.

de Meeûs, T. and Renaud, F. 2002. Parasites within the new phylogeny of eukaryotes. *Trends in Parasitology* 18: 247–251.

Desdevises, Y., Morand, S., Jousson, O. and Legendre, P. 2002. Coevolution between *Lamellodiscus* (Monogenea: Diplectanidae) and Sparidae (Teleostei): the study of a complex host-parasite system. *Evolution* 56: 2459–2471.

Desdevises, Y., Morand, S. and Oliver, G. 2001. Linking specialisation to diversification in the Diplectanidae Bychowsky 1957 (Monogenea, Platyhelminthes). *Parasitology Research* 87: 223–230.

Dezfuli, B.S., Giari, L., De Biaggi, S. and Poulin, R. 2001. Associations and interactions among intestinal helminths of the brown trout, *Salmo trutta*, in northern Italy. *Journal of Helminthology* 75: 331–336.

Dial, K.P. and Marzluff, J.M. 1988. Are the smallest organisms the most diverse? *Ecology* 69: 1620–1624.

Dial, K.P. and Marzluff, J.M. 1989. Nonrandom diversification within taxonomic assemblages. *Systematic Zoology* 38: 26–37.

Diaz-Uriarte, R. and Garland, T. Jr. 1998. Effects of branch length errors on the performance of phylogenetically independent contrasts. *Systematic Biology* 47: 654–672.

Dieckmann, U. and Doebeli, M. 1999. On the origin of species by sympatric speciation. *Nature* 400: 354–357.

Diekmann, O. and Heesterbeek, J.A.P. 2000. *Mathematical Epidemiology of Infectious Diseases: Model Building, Analysis and Interpretation.* John Wiley and Sons, Chichester, NY.

Dobson, A.P. 1989. The population biology of parasitic helminths in animal populations. In: *Applied Mathematical Ecology* (Levin, S.A., Hallam, T.G. and Gross, L.J., editors). Springer-Verlag, Berlin, pp. 145–175.

Dobson, A.P. 1990. Models for multi-species parasite-host communities. In: *Parasite Communities: Patterns and Processes* (Esch, G.W., Bush, A.O. and Aho, J.M., editors). Chapman and Hall, London, pp. 261–288.

Dobson, A.P. and Merenlender, A. 1991. Coevolution of macroparasites and their hosts. In: *Parasite-Host Associations: Coexistence or Conflict?* (Toft, C.A., Aeschlimann, A. and Bolis, L., editors). Oxford University Press, Oxford, pp. 83–101.

Dobson, A.P. and Pacala, S.W. 1992. The parasites of *Anolis* lizards in the northern Lesser Antilles. 2. The structure of the parasite community. *Oecologia* 91: 118–125.

Dobson, A.P., Pacala, S.W., Roughgarden, J.D., Carper, E.R. and Harrics, E.A. 1992. The parasites of *Anolis* lizards in the northern Lesser Antilles. 1. Patterns of distribution and abundance. *Oecologia* 91: 110–117.

Dobson, A.P. and Roberts, M. 1994. The population dynamics of parasitic helminth communities. *Parasitology* 109: S97–S108.

Dolphin, K. and Quicke, D.L.J. 2001. Estimating the global species richness of an incompletely described taxon: an example using parasitoid wasps (Hymenoptera: Braconidae). *Biological Journal of the Linnean Society* 73: 279–286.

Dove, A.D.M. 1998. A silent tragedy: parasites and the exotic fishes of Australia. *Proceedings of the Royal Society of Queensland* 107: 109–113.

Dove, A.D.M. and Fletcher, A.S. 2000. The distribution of the introduced tapeworm *Bothriocephalus acheilognathi* in Australian freshwater fishes. *Journal of Helminthology* 74: 121–127.

Dowling, A.P.G. 2002. Testing the accuracy of TreeMap and Brooks parsimony analyses of coevolutionary patterns using artificial associations. *Cladistics* 18: 416–435.

Dowling, A.P.G, van Veller, M.G.P., Hoberg, E.P and Brooks, D.R. 2003. A priori and a posteriori methods in comparative evolutionary studies of host-parasite associations. *Cladistics* 19: 240–253.

Dritschilo, W., Cornell, H., Nafus, D. and O'Connor, B. 1975. Insular biogeography: of mice and mites. *Science* 190: 467–469.

Dupas, S., Morand, S. and Eslin, P. 2004. Evolution of hemocyte concentration in the *melanogaster* subgroup species. *Comptes Rendus de l'Académie des Sciences, Biologie* (in press).

Du Pasquier, L. 1982. Antibody diversity in lower vertebrates: why is it so restricted? *Nature* 296: 311–313.

Dybdahl, M.F. and Lively, C.M. 1998. Host-parasite coevolution: evidence for rare advantage and time-lagged selection in a natural population. *Evolution* 52: 1057–1066.

Ebert, D., Hottinger, J.W. and Pajunen, V.I. 2001. Temporal and spatial dynamics of parasite richness in a *Daphnia* metapopulation. *Ecology* 82: 3417–3434.

Egid, K. and Brown, J.L. 1989. The major histocompatibility complex and female mating preferences in mice. *Animal Behaviour* 38: 548–550.

Elgar, M.A. and Harvey, P.H. 1987. Basal metabolic rates in mammals: allometry, phylogeny and ecology. *Functional Ecology* 1: 25–36.

Ellegren, H., Mikko, S., Wallin, K. and Andersson, L. 1996. Limited polymorphism at major histocompatibility complex (MHC) loci in the Swedish moose A. *alces*. *Molecular Ecology* 5: 3–9.

Enquist, B.J., Haskell, J.P. and Tiffney, B.H. 2002. General patterns of taxonomic and biomass partitioning in extant and fossil plant communities. *Nature* 419: 610–613.

Erwin, T.L. 1982. Tropical forests: their richness in Coleoptera and other arthropod species. *The Coleopterist's Bulletin* 36: 74–75.

Esch, G.W., Curtis, L.A. and Barger, M.A. 2001. A perspective on the ecology of trematode communities in snails. *Parasitology* 123: S57–75.

Esch, G.W., Kennedy, C.R., Bush, A.O. and Aho, J.M. 1988. Patterns in helminth communities in freshwater fish in Great Britain: alternative strategies for colonization. *Parasitology* 96: 519–532.

Euzet, L., Renaud, F. and Gabrion, C. 1984. Le complexe *Bothriocephalus scorpii* (Mueller, 1776): différenciation à l'aide des méthodes biochimiques de deux espèces parasites du turbot (*Psetta maxima*) et de la barbue (*Scophthalmus rhombus*). *Bulletin de la Société Zoologique de France* 109: 84–88.

Ezenwa, V.O. 2003. Habitat overlap and gastrointestinal parasitism in sympatric African bovids. *Parasitology* 126: 379–388.

Fair, M., Hansen, E.S and Ricklefs, R.E. 1999. Growth, developmental stability and immune response in juvenile Japanese quails (*Coturnix coturnix japonica*). *Proceedings of the Royal Society of London* B 266: 1735–1742.

Faivre, B., Grégoire, A., Préault, M., Cézilly, F. and Sorci, G. 2003. Immune activation rapidly mirrored in a secondary sexual trait in blackbirds. *Science* 300: 103.

Fan, W., Liu, Y.-C., Parimoo, S. and Weissman, S.M. 1995. Olfactory receptor-like genes are located in the human major histocompatibility complex. *Genomics* 27: 119–123.

Farrell, B.D., Mitter, C. and Futuyma, D.J. 1992. Diversification at the insect-plant interface. *BioScience* 42: 34–42.

Feliu, C., Renaud, F., Catzeflis, F., Durand, P., Hugot, J.-P. and Morand, S. 1997. A comparative analysis of parasite species richness of Iberian rodents. Parasitology 115: 453–466.

Feliu, C., Spakulova, M., Casanova, J.C., Renaud, F., Morand, S., Hugot, J.-P., Santanella, F. and Durand, P. 2000. Genetic and morphological heterogeneity in small rodent whipworms in southwestern Europe: characterization of *Trichuris muris* and description of *Trichuris arvicolae* n. sp. (Nematoda: Trichuridae). *Journal of Parasitology* 86: 442–449.

Felsenstein, J. 1985. Phylogenies and the comparative method. *American Naturalist* 125: 1–15.

Fenchel, T. 1993. There are more small than large species? *Oikos* 68: 375–378.

Fletcher, A.S. and Whittington, I.D. 1998. A parasite-host checklist for Monogenea from freshwater fishes in Australia, with comments on biodiversity. *Systematic Parasitology* 41: 159–168.

Folstad, I. and Karter, A.J. 1992. Parasites, bright males, and the immunocompetence handicap. *American Naturalist* 139: 603–622.

Font, W.F. 1998. Parasites in paradise: patterns of helminth distribution in Hawaiian stream fishes. *Journal of Helminthology* 72: 307–311.

Font, W.F. 2003. The global spread of parasites: what do Hawaiian streams tell us? *BioScience* 53: 1061–1067.

Font, W.F. and Tate, D.C. 1994. Helminth parasites of native Hawaiian freshwater fishes: an example of extreme ecological isolation. *Journal of Parasitology* 80: 682–688.

Forrester, D.J., Conti, J.A. and Belden, R.C. 1985. Parasites of the Florida panther (*Felis concolor coryi*). *Proceedings of the Helminthological Society of Washington* 52: 95–97.

Frank, S.A. 1996. Models of parasite virulence. *Quarterly Review of Biology* 71: 37–78.

Freeland, W.J. 1979. Primate social groups as biological islands. *Ecology* 60: 719–728.

Fromont, E., Morvilliers, L., Artois, M. and Pontier, D. 2001. Parasite richness and abundance in insular and mainland feral cats: insularity or density? *Parasitology* 123: 143–151.

Frost, S.D.W. 1999. The immune system as an inducible defense. In: *The Ecology and Evolution of Inducible Defenses* (Tollrian, R. and Harvell, C.D., editors). Princeton University Press, Princeton, pp. 104–126.

Futuyma, D.J. and Moreno, G. 1988. The evolution of ecological specialization. *Annual Review of Ecology and Systematics* 19: 207–233.

Gage, J.D. 1996. Why are there so many species in deep-sea sediments? *Journal of Experimental Marine Biology and Ecology* 200: 257–286.

Gardezi, T. and da Silva, J. 1999. Diversity in relation to body size in mammals: a comparative study. *American Naturalist* 153: 110–123.

Gartner, J.V.Jr. and Zwerner, D.E. 1989. The parasite faunas of meso- and bathypelagic fishes of Norfolk Submarine Canyon, western North Atlantic. *Journal of Fish Biology* 34: 79–95.

Gaston, K.J. 1991a. Body size and probability of description: the beetle fauna of Britain. *Ecological Entomology* 16: 505–508.

Gaston, K.J. 1991b. The magnitude of global insect species richness. *Conservation Biology* 5: 283–296.

Gaston, K.J. 1996. Spatial covariance in the species richness of higher taxa. In: *Aspects of the Genesis and Maintenance of Biological Diversity* (Hochberg, M.E., Clobert, J. and Barbault, R., editors). Oxford University Press, Oxford, pp. 221–242.

Gaston, K.J. and Blackburn, T.M. 1994. Are newly described bird species small-bodied? *Biodiversity Letters* 2: 16–20.

Gaston, K.J. and Blackburn, T.M. 2000. *Pattern and Process in Macroecology.* Blackwell Science, Oxford.

Gaston, K.J., Blackburn, T.M. and Loder, N. 1995. Which species are described first? The case of North American butterflies. *Biodiversity and Conservation* 4: 119–127.

Gaston, K.J. and Hudson, E. 1994. Regional patterns of diversity and estimates of global insect species richness. *Biodiversity and Conservation* 3: 493–500.

Gibson, D.I. and Bray, R.A. 1994. The evolutionary expansion and host-parasite relationships of the Digenea. *International Journal for Parasitology* 24: 1213–1226.

Gittleman, J.L. and Purvis, A. 1998. Body size and species richness in carnivores and primates. *Proceedings of the Royal Society of London* B 265: 113–119.

Glaw, F. and Köhler, J. 1998. Amphibian species diversity exceeds that of mammals. *Herpetological Reviews* 29: 11–12.

Goater, C.P. and Holmes, J.C. 1997. Parasite-mediated natural selection. In: *Host-Parasite Evolution: General Principles and Avian Models* (Clayton, D.H. and Moore, J., editors). Oxford University Press, Oxford, pp. 9–29.

Goldberg, S.R. and Bursey, C.R. 2000. Helminths of Mexican lizards: geographical distribution. In: *Metazoan Parasites in the Neotropics: A Systematic and Ecological Perspective* (Salgado-Maldonado, G., Aldrete, A.N.G. and Vidal-Martinez, V.M., editors). Universidad Nacional Autónoma de México, Mexico, pp. 175–191.

Golvan, Y.J. 1994. Nomenclature of the Acanthocephala. *Research and Reviews in Parasitology* 54: 135–205.

Gonzalez, G., Sorci, G., Møller, A.P., Ninni, P., Hausey, C. and De Lope, F. 1999.

Immunocompetence and condition-dependent sexual advertisement in male house sparrows (*Passer domesticus*). *Journal of Animal Ecology* 68: 1225–1234.

Goüy de Bellocq, J., Morand, S. and Feliu, C. 2002. Patterns of parasite species richness of Western Palaearctic micro-mammals: island effects. *Ecography* 25: 173–183.

Goüy de Bellocq, J., Sara, M., Casanova, J.-C., Feliu, C. and Morand, S. 2003. A comparison of the structure of helminth communities in the wood mouse, *Apodemus sylvaticus*, on islands of the western Mediterranean and continental Europe. *Parasitology Research* 90: 64–70.

Gray, J.S. 1994. Is deep-sea species diversity really so high? Species diversity of the Norwegian continental shelf. *Marine Ecology Progress Series* 112: 205–209.

Gregory, R.D. 1990. Parasites and host geographic range as illustrated by waterfowl. *Functional Ecology* 4: 645–654.

Gregory, R.D., Keymer, A.E. and Harvey, P.H. 1991. Life history, ecology and parasite community structure in Soviet birds. *Biological Journal of the Linnean Society* 43: 249–262.

Gregory, R.D., Keymer, A.E. and Harvey, P.H. 1996. Helminth parasite richness among vertebrates. *Biodiversity and Conservation* 5: 985–997.

Grenfell, B.T. and Dobson, A.P. 1995. *Ecology of Infectious Diseases in Natural Populations*. Cambridge University Press, Cambridge.

Groombridge, B. 1992. *Global Diversity: Status of the Earth's Living Resources*. Chapman and Hall, London.

Grossman, C.J. 1985. Interactions between the gonadal steroids and the immune system. *Science* 227: 257–261.

Grygier, M.J. 1987. Classification of the Ascothoracida (Crustacea). *Proceedings of the Biological Society of Washington* 100: 452–458.

Guégan, J.-F. and Kennedy, C.R. 1993. Maximum local helminth parasite community richness in British freshwater fish: a test of the colonization time hypothesis. *Parasitology* 106: 91–100.

Guégan, J.-F. and Kennedy, C.R. 1996. Parasite richness / sampling effort / host range: the fancy three-piece jigsaw puzzle. *Parasitology Today* 12: 367–369.

Guégan, J.-F., Lambert, A., Lévêque, C., Combes, C. and Euzet, L. 1992. Can host body size explain the parasite species richness in tropical freshwater fishes? *Oecologia* 90: 197–204.

Guégan, J.-F. and Morand, S. 1996. Polyploid hosts: strange attractors for parasites? *Oikos* 77: 366–370.

Gunnarsson, B. 1992. Fractal dimension of plants and body size distributions in spiders. *Functional Ecology* 6: 636–641.

Hafner, M.S. and Nadler, S.A. 1988. Phylogenetic trees support the coevolution of parasites and their hosts. *Nature* 332: 258–259.

Hafner, M.S. and Nadler, S.A. 1990. Cospeciation in host-parasite assemblages: comparative analysis of rates of evolution and timing of cospeciation events. *Systematic Zoology* 39: 192–204.

Halmetoja, A., Valtonen, E.T. and Koskenniemi, E. 2000. Perch (*Perca fluviatilis* L.) parasites reflect ecosystem conditions: a comparison of a natural lake and two acidic reservoirs in Finland. *International Journal for Parasitology* 30: 1437–1444.

Hamilton, W.D. 1980. Sex versus non-sex versus parasite. *Oikos* 35: 282–290.

Hamilton, W.D. 1982. Pathogens as causes of genetic diversity in their host populations. In: *Population Biology of Infectious Diseases* (Anderson, R.M. and May, R.M., editors). Springer, New York, pp. 269–296.

Hamilton, W.D., Axelrod, R. and Tanese, R. 1990. Sexual reproduction as an adaptation to resist parasites. *Proceedings of the National Academy of Sciences U.S.A.* 87: 3566–3573.

Hamilton, W.D. and Zuk, M. 1982. Heritable true fitness and bright birds: a role for parasites? *Science* 218: 384–386.

Hamilton, W.J. and Poulin, R. 1997. The Hamilton and Zuk hypothesis revisited: a meta-analytical approach. *Behaviour* 134: 299–320.

Hammond, P.M. 1992. Species inventory. In: *Global Diversity: Status of the Earth's Living Resources* (Groombridge, B., editor). Chapman and Hall, London, pp. 17–39.

Hanken, J. 1999. Why are there so many new amphibian species when amphibians are declining? *Trends in Ecology and Evolution* 14: 7–8.

Hanley, K.A., Fisher, R.N. and Case, T.J. 1995. Lower mite infestations in an asexual gecko compared with its sexual ancestors. *Evolution* 49: 418–426.

Hanski, I.A. and Simberloff, D. 1997. The metapopulation approach, its history, conceptual domain, and application to conservation. In: *Metapopulation Biology: Ecology, Genetics, and Evolution* (Hanski, I.A. and Gilpin, M.E., editors). Academic Press, San Diego, pp. 5–26.

Hartvigsen, R. and Halvorsen, O. 1993. Common and rare trout parasites in a small landscape system. *Parasitology* 106: 101–105.

Hartvigsen, R. and Halvorsen, O. 1994. Spatial patterns in the abundance and distribution of parasites of freshwater fish. *Parasitology Today* 10: 28–31.

Harvey, P.H. and Pagel, M.D. 1991. *The Comparative Method in Evolutionary Biology*. Oxford University Press, Oxford.

Harvey, P.H., Pagel, M.D. and Rees, J.A. 1991. Mammalian metabolism and life histories. *American Naturalist* 137: 556–566.

Hasselquist, D., Wasson, M.F. and Winkler, D.W. 2001. Humoral immunocompetence correlates with date of egg-laying and reflects work load in female tree swallows. *Behavioral Ecology* 12: 93–97.

Haukisalmi, V. and Henttonen, H. 1993. Coexistence in helminths of the bank vole *Clethrionomys glareolus*. I. Patterns of co-occurrence. *Journal of Animal Ecology* 62: 221–229.

Hawkins, B.A. 1994. *Pattern and Process in Host-Parasitoid Interactions*. Cambridge University Press, Cambridge.

Hayward, C.J. 1997. Distribution of external parasites indicate boundaries to dispersal of sillaginid fishes in the Indo-West Pacific. *Marine and Freshwater Research* 48: 391–400.

He, F. and Legendre, P. 1996. On species-area relations. *American Naturalist* 148: 719–737.

Hedrick, P.W. 1994. Evolutionary genetics of the major histocompatibility complex. *American Naturalist* 143: 945–964.

Heltshe, J.F. and Forrester, N.E. 1983. Estimating species richness using the jackknife procedure. *Biometrics* 39: 1–12.

Hengeveld, R. 1990. *Dynamic Biogeography*. Cambridge University Press, Cambridge.

Hernández-Alcántara, P. and Solis-Weiss, V. 1998. Parasitism among polychaetes: a rare case illustrated by a new species: *Labrorostratus zaragozensis*, n. sp. (Oenonidae) found in the Gulf of California, Mexico. *Journal of Parasitology* 84: 978–982.

Heywood, V.H., Mace, G.M., May, R.M. and Stuart, S.N. 1994. Uncertainties in extinction rates. *Nature* 368: 105.

Hill, A.V.S., Allsopp, C.E.M. and Kwiatowski, D. 1991. Common West African HLA antigens associated with protection from severe malaria. *Nature* 352: 595–600.

Hillebrand, H. and Azovsky, A.I. 2001. Body size determines the strength of the latitudinal diversity gradient. *Ecography* 24: 251–256.

Hillgarth, N. and Wingfield, J.C. 1997. Testosterone and immunosuppression in vertebrates: implications for parasite-mediated sexual selection. In: *Parasites and Pathogens: Effects on Host Hormones and Behavior* (Beckage, N.E., editor). Chapman and Hall, New York, pp. 143–155.

Hillis, D.M., Moritz, C. and Mable, B.K. 1996. *Molecular Systematics*, second edition. Sinauer, Sunderland, MA.

Hoberg, E.P. 1992. Congruent and synchronic patterns in biogeography and speciation among seabirds, pinnipeds and cestodes. *Journal of Parasitology* 78: 601–615.

Hoberg, E.P. 1995. Historical biogeography and modes of speciation across high latitude seas of the Holarctic: concepts for host-parasite coevolution among Phocini (Phocidae) and Tetrabothriidae (Eucestoda). *Canadian Journal of Zoology* 73: 45–57.

Hoberg, E.P. 1996. Faunal diversity among avian parasite assemblages: the interaction of history, ecology, and biogeography in marine systems. *Bulletin of the Scandinavian Society of Parasitology* 6: 65–89.

Hoberg, E.P. 1997. Phylogeny and historical reconstruction: host-parasite systems as keystones in biogeography and ecology. In: *Biodiversity II: Understanding and Protecting our Biological Resources* (Reaka-Kudla, M.L., Wilson, D.E. and Wilson, E.O., editors). Joseph Henry Press, Washington, pp. 243–261.

Hoberg, E.P., Brooks, D.R. and Siegel-Causey, D. 1997. Host-parasite cospeciation: history, principles and prospects. In: *Host-Parasite Evolution: General Principles and Avian Models* (Clayton, D.H. and Moore, J., editors). Oxford University Press, Oxford, pp. 212–235.

Hoberg, E.P. and Klassen, G.J. 2002. Revealing the faunal tapestry: coevolution and historical biogeography of hosts and parasites in marine systems. *Parasitology* 124: S3–S22.

Hoberg, E.P., Monsen, K.J., Kutz, S. and Blouin, M.S. 1999. Structure, biodiversity, and historical biogeography of nematode faunas in holarctic ruminants: morphological and molecular diagnoses for *Teladorsagia boreoarcticus* n. sp. (Nematoda: Ostertagiinae), a dimorphic cryptic species in muskoxen (*Ovibos moschatus*). *Journal of Parasitology* 85: 910–934.

Hochberg, M.E., Michalakis, Y. and de Meeus, T. 1992. Parasitism as a constraint on the rate of life-history evolution. *Journal of Evolutionary Biology* 5: 491–504.

Høeg, J.T. 1995. The biology and life cycle of the Rhizocephala (Cirripedia). *Journal of the Marine Biological Association U.K.* 75: 517–550.

Holland, C. 1984. Interactions between *Moniliformis* (Acanthocephala) and *Nippostrongylus* (Nematoda) in the small intestine of laboratory rats. *Parasitology* 88: 303–315.

Holmes, J.C. 1961. Effects of concurrent infections on *Hymenolepis diminuta* (Cestoda) and *Moniliformis dubius* (Acanthocephala). I. General effects and comparison with crowding. *Journal of Parasitology* 47: 209–216.

Holmes, J.C. and Price, P.W. 1986. Communities of parasites. In: *Community Ecology: Pattern and Process* (Anderson, D.J. and Kikkawa, J., editors). Blackwell Scientific Publications, Oxford, pp. 187–213.

Holmstad, P.R. and Skorping, A. 1998. Covariation of parasite intensities in willow ptarmigan, *Lagopus lagopus* L. *Canadian Journal of Zoology* 76: 1581–1588.

Horak, P., Ots, I. and Murumägi, A. 1998. Hematological health state indices of reproducing great tits: a response to brood size manipulation. *Functional Ecology* 12: 750–736.

Horak P., Tegelmann, L., Ots, I. and Møller, A.P. 1999. Immune function and survival of great tit nestlings in relation to growth conditions. *Oecologia* 121: 316–322.

Hou, Y., Suzuki, Y. and Aida, K. 1999 Effects of steroid hormones on immunoglobulin M (IgM) in rainbow trout. *Fish Physiology and Biochemistry* 20: 155–162.

Houck, M.A. 1994. *Mites: Ecological and Evolutionary Analyses of Life-History Patterns.* Chapman and Hall, New York.

Hubbell, S.P. 2001. *The Unified Neutral Theory of Biodiversity and Biogeography.* Princeton University Press, Princeton.

Hudson, P.J., Dobson, A.P. and Newborn, D. 1998. Prevention of population cycles by parasite removal. *Science* 282: 2256–2258.

Hudson, P.J. and Greenman, J. 1998. Competition mediated by parasites: biological and theoretical progress. *Trends in Ecology and Evolution* 13: 387 390.

Hudson, P.J., Rizzoli, A., Grenfell, B.T., Heesterbeek, H. and Dobson, A.P. 2002. *The Ecology of Wildlife Diseases*. Oxford University Press, Oxford.

Hughes, A.L. and Nei, M. 1989. Nucleotide substitution at major histocompatibility complex class II loci: evidence for overdominant selection. *Proceedings of the National Academy of Sciences U.S.A.* 86: 958–962.

Hughes, A.L. and Yeager, M. 1998. Natural selection at major histocompatibility complex. *Annual Review of Genetics* 32: 415–435.

Hugot, J.-P., Baujard, P. and Morand, S. 2001. Biodiversity in helminths and nematodes as a field of study: an overview. *Nematology* 3: 1–10.

Humes, A.G. 1994. How many copepods? *Hydrobiologia* 292/293: 1–7.

Hung, G.C., Chilton, N.B., Beveridge, I., Zhu, X.Q., Lichtenfels, J.R. and Gasser, R.B. 1999. Molecular evidence for cryptic species within *Cylicostephanus minutus* (Nematoda: Strongylidae). *International Journal for Parasitology* 29: 285–291.

Huston, M.A. 1992. *Biological Diversity: The Coexistence of Species on Changing Landscapes*. Cambridge University Press, Cambridge.

Hutchinson, G.E. 1959. Homage to Santa Rosalia or why are there so many kinds of animal species? *American Naturalist* 93: 145–159.

Huxham, M., Raffaelli, D. and Pike, A. 1995. Parasites and food web patterns. *Journal of Animal Ecology* 64: 168–176.

Huyse, T. and Volckaert, F.A.M. 2002. Identification of a host-associated species complex using molecular and morphometric analyses, with the description of *Gyrodactylus rugiensoides* n. sp. (Gyrodactylidae, Monogenea). *International Journal for Parasitology* 32: 907–919.

Inglis, W.G. 1971. Speciation in parasitic nematodes. *Advances in Parasitology* 9: 201–223.

Isaac, N.J.B., Agapow, P.-M., Harvey, P.H. and Purvis, A. 2003. Phylogenetically nested comparisons for testing correlates of species richness: a simulation study of continuous variables. *Evolution* 57: 18–26.

Isaacson, P.J. and Webster, J.M. 2002. Antimicrobial activity of *Xenorhabus* sp RIO (Enterobacteriaceae), symbiont of the entomopathogenic nematode, *Steinernema riobrave* (Rhabditida: Steinernematidae). *Journal of Invertebrate Pathology* 79: 146–153.

Ives, A.R. 1991. Aggregation and coexistence in a carrion fly community. *Ecological Monographs* 61: 75–94.

Jablonski, D. 1986. Background and mass extinctions: the alternation of macroevolutionary regimes. *Science* 231: 129–133.

Jaenike, J. 1978. An hypothesis to account for the maintenance of sex within populations. *Evolutionary Theory* 3: 191–194.

Jaenike, J. and James, A.C. 1991. Aggregation and the coexistence of mycophagous *Drosophila*. *Journal of Animal Ecology* 60: 913–928.

John, J.L. 1994. Nematodes and the spleen; an immunological relationship. *Experientia* 50: 15–22.

John, J.L. 1995. Parasites and the avian spleen. *Biological Journal of the Linnean Society* 54: 87–106.

Johnson, K.P., Adams, R.J. and Clayton, D.H. 2002a. The phylogeny of the louse genus *Brueelia* does not reflect host phylogeny. *Biological Journal of the Linnean Society* 77: 233–247.

Johnson, K.P., Weckstein, J.D., Witt, C.C., Faucett, R.C. and Moyle, R.G. 2002b. The perils of using host relationships in parasite taxonomy: phylogeny of the *Degeeriella* complex. *Molecular Phylogenetics and Evolution* 23: 150–157.

Jong, S.C., Birmingham, J.M. and Pai, S.H. 1991. Immunomodulatory substances of fungal origin. *Rivista di Immunologia ed Immunofarmacologia* 11: 115–122.

Jonsson, B.G. 2001. A null model for randomization tests of nestedness in species assemblages. *Oecologia* 127: 309–313.

Jordan, W.C. and Bruford, M.W. 1998. New perspectives on mate choice and the MHC diversity. *Heredity* 81: 239–245.

Jousson, O., Bartoli, P. and Pawlowski, J. 2000. Cryptic speciation among intestinal parasites (Trematoda: Digenea) infecting sympatric host fishes (Sparidae). *Journal of Evolutionary Biology* 13: 778–785.

Kabata, Z. 1979. *Parasitic Copepoda of the British Isles*. The Ray Society, London.

Kearn, G.C. 1998. *Parasitism and the Platyhelminths*. Chapman and Hall, London.

Keas, B.E. and Blankespoor, H.D. 1997. The prevalence of cercariae from *Stagnicola emarginata* (Lymnaeidae) over 50 years in northern Michigan. *Journal of Parasitology* 83: 536–540.

Kelly, C.K. and Southwood, T.R.E. 1999. Species richness and resource availability: a phylogenetic analysis of insects associated with trees. *Proceedings of the National Academy of Sciences U.S.A.* 96: 8013–8016.

Kennedy, C.E.J. and Southwood, T.R.E. 1984. The number of species of insect associated with British trees: a reanalysis. *Journal of Animal Ecology* 53: 455–478.

Kennedy, C.R. 1978. An analysis of the metazoan parasitocoenoses of brown trout *Salmo trutta* from British lakes. *Journal of Fish Biology* 13: 255–263.

Kennedy, C.R. 1995. Richness and diversity of macroparasite communities in eels *Anguilla reinhardtii* in Queensland, Australia. *Parasitology* 111: 233–245.

Kennedy, C.R. and Bush, A.O. 1992. Species richness in helminth communities: the importance of multiple congeners. *Parasitology* 104: 189–197.

Kennedy, C.R. and Bush, A.O. 1994. The relationship between pattern and scale in parasite communities: a stranger in a strange land. *Parasitology* 109: 187–196.

Kennedy, C.R., Bush, A.O. and Aho, J.M. 1986. Patterns in helminth communities: why are birds and fish different? *Parasitology* 93: 205–215.

Kennedy, C.R. and Guégan, J.-F. 1994. Regional versus local helminth parasite richness in British freshwater fish: saturated or unsaturated parasite communities? *Parasitology* 109: 175–185.

Kennedy, C.R. and Guégan, J.-F. 1996. The number of niches in intestinal helminth communities of *Anguilla anguilla*: are there enough spaces for parasites? *Parasitology* 113: 293–302.

Khan, R.A. and Thulin, J. 1991. Influence of pollution on parasites of aquatic animals. *Advances in Parasitology* 30: 201–238.

Kiesecker, J.M. and Skelly, D.K. 2000. Choice of oviposition site by gray treefrogs: the role of potential parasitic infection. *Ecology* 81: 2939–2943.

Kim, C.B. and Kim, W. 1993. Phylogenetic relationships among gammaridean families and amphipod suborders. *Journal of Natural History* 27: 933–946.

Kim, K.C. 1985. *Coevolution of Parasitic Arthropods and Mammals*. John Wiley and Sons, New York.

Kinsella, J.M. 1971. Growth, development, and intraspecific variation of *Quinqueserialis quinqueserialis* (Trematoda: Notocotylidae) in rodent hosts. *Journal of Parasitology* 57: 62–70.

Klassen, G.J. 1992. Coevolution: a history of the macroevolutionary approach to studying host-parasite associations. *Journal of Parasitology* 78: 573–587.

Klein, J. 1986. *Natural History of the Major Histocompatibility Complex*. Wiley, New York.

Klein, J. 1991. Of HLA, Tryps, and selection: an essay on coevolution of MHC and parasites. *Human Immunology* 30: 247–258.

Klompen, J.S.H., Black, W.C., Keirans, J.E. and Oliver, J,H. 1996. Evolution of ticks. *Annual Review of Entomology* 41: 141–161.

Koella, J.C. 2000. Coevolution of parasite life cycles and host life-histories. In: *Evolutionary Biology of Host-Parasite Relationships: Theory Meets Reality* (Poulin, R., Morand, S. and Skorping, A., editors). Elsevier, Amsterdam, pp. 185–200.

Kondrashov, A.S. and Kondrashov, F.A. 1999. Interactions among quantitative traits in the course of sympatric speciation. *Nature* 400: 351–354.

Koteja, P. 1991. On the relation between basal and field metabolic rates in birds and mammals. *Functional Ecology* 5: 56–64.

Kozlowski, J. 2002. Theoretical and empirical status of Brown, Marquet and Taper's model of species-size distribution. *Functional Ecology* 16: 540–542.

Kozlowski, J. and Gawelczyk, A.T. 2002. Why are species' body size distributions usually skewed to the right? *Functional Ecology* 16: 419–432.

Krasnov, B.R., Shenbrot, G.I., Medvedev, S.G., Vatschenok, V.S. and Khokhlova, I.S. 1997. Host-habitat relations as an important determinant of spatial distribution of flea assemblages (Siphonaptera) on rodents in the Negev Desert. *Parasitology* 114: 159–173.

Krecek, R.C., Malan, F.S., Rupprecht, C.E. and Childs, J.E. 1987. Nematode parasites from Burchell's zebras in South Africa. *Journal of Wildlife Diseases* 23: 404–411.

Krist, A.C. 2001. Variation in fecundity among populations of snails is predicted by prevalence of castrating parasites. *Evolutionary Ecology Research* 3: 191–197.

Kunz, W. 2002. When is a parasite species a species? *Trends in Parasitology* 18: 121–124.

Kuris, A.M. and Blaustein, A.R. 1977. Ectoparasitic mites on rodents: application of the island biogeography theory? *Science* 195: 596–598.

Kuris, A.M., Blaustein, A.R. and Alió, J.J. 1980. Hosts as islands. *American Naturalist* 116: 570–586.

Kuris, A.M. and Lafferty, K.D. 1994. Community structure: larval trematodes in snail hosts. *Annual Review of Ecology and Systematics* 25: 189–217.

Lafferty, K.D. 1993. The marine snail, *Cerithidea californica,* matures at smaller sizes where parasitism is high. *Oikos* 68: 3–11.

Lafferty, K.D. 1997. Environmental parasitology: What can parasites tell us about human impacts on the environment? *Parasitology Today* 13: 251–255.

Langerfors, A., Lohm, J., Grahn, M., Andersen, O. and von Schantz T. 2001. Association between major Histocompatibility complex IIB and resistance to *Aeromonas salmonicida. Proceedings of the Royal Society of London* B 268: 479–485.

Lawton, J.H. 1983. Plant architecture and the diversity of phytophagous insects. *Annual Review of Entomology* 28: 23–29.

Lawton, J.H. 1999. Are there general laws in ecology? *Oikos* 84: 177–192.

Leignel, V., Cabaret, J. and Humbert, J.F. 2002. New molecular evidence that *Teladorsagia circumcincta* (Nematoda: Trichostrongylidea) is a species complex. *Journal of Parasitology* 88: 135–140.

Levsen, A. 2001. Transmission ecology and larval behaviour of *Camallanus cotti* (Nematoda, Camallanidae) under aquarium conditions. *Aquarium Sciences and Conservation* 3: 315–325.

Levsen, A. and Jakobsen, P.J. 2002. Selection pressure towards monoxeny in *Camallanus cotti* (Nematoda, Camallanidae) facing an intermediate host bottleneck situation. *Parasitology* 124: 625–629.

Liersch, S. and Schmid-Hempel, P. 1998. Genetic variation within social insect colonies reduces parasite load. *Proceedings of the Royal Society of London* B 265: 221–225.

Lim, L.H.S. 1998. Diversity of monogeneans in Southeast Asia. *International Journal for Parasitology* 28: 1495–1515.

Littlewood, D.T.J., Rohde, K. and Clough, K.A. 1997. Parasite speciation within or between host species? Phylogenetic evidence from site-specific polystome monogeneans. *International Journal for Parasitology* 27: 1289–1297.

Lively, C.M. 1987. Evidence from a New Zealand snail for the maintenance of sex by parasitism. *Nature* 328: 519–521.

Lively, C.M., Craddock, C. and Vrijenhoek, R.C. 1990. Red queen hypothesis supported by parasitism in sexual and clonal fish. *Nature* 344: 864–866.

Lo, C.M., Morgan, J.A.T., Galzin, R. and Cribb, T.H. 2001. Identical digeneans in coral reef fishes from French Polynesia and the Great Barrier Reef (Australia) demonstrated by morphology and molecules. *International Journal for Parasitology* 31: 1573–1578.

Lochmiller, R.L. and Deerenberg, C. 2000. Trade-offs in evolutionary immunology: just what is the cost of immunity? *Oikos* 88: 87–98.

Lotz, J.M., Bush, A.O. and Font, W.F. 1995. Recruitment-driven, spatially discontinuous communities: a null model for transferred patterns in target communities of intestinal helminths. *Journal of Parasitology* 81: 12–24.

Lotz, J.M. and Font, W.F. 1991. The role of positive and negative interspecific associations in the organization of communities of intestinal helminths of bats. *Parasitology* 103: 127–138.

Lotz, J.M. and Font, W.F. 1994. Excess positive associations in communities of intestinal helminths of bats: a refined null hypothesis and a test of the facilitation hypothesis. *Journal of Parasitology* 80: 398–413.

Lowenberger, C.A. and Rau, M.E. 1994. Selective oviposition by *Aedes aegypti* (Diptera: Culicidae) in response to a larval parasite, *Plagiorchis elegans* (Trematoda: Plagiorchiidae). *Environmental Entomology* 23: 1269–1276.

Luque, J.L., Mouillot, D. and Poulin, R. 2004. Parasite biodiversity and its determinants in coastal marine teleost fishes of Brazil. *Parasitology* 128: 671–682.

MacArthur, R.H. and Wilson, E.O. 1967. *The Theory of Island Biogeography*. Princeton University Press, Princeton.

MacDonald, G.M. 2001. *Biogeography: An Introduction to Space, Time and Life*. Wiley, Chichester, NY.

MacManus, D.P. and Bowles, J. 1996. Molecular genetic approaches to parasite identification: their value in diagnostic parasitology and systematics. *International Journal for Parasitology* 26: 687–704.

Macpherson, E. 2002. Large-scale species richness gradients in the Atlantic Ocean. *Proceedings of the Royal Society of London* B 269: 1715–1720.

Magurran, A.E. 1988. *Ecological Diversity and its Measurement*. Croom Helm, London.

Maier, S.F. and Watkins, L.R. 1999. Bidirectional communications between behavior, brain and immunology: implications for behaviour. *Animal Behaviour* 57: 741–751.

Maier, S.F., Watkins, L.R. and Fleshner, M. 1994. Psychoneuroimmunology: the interface between behavior, brain and immunology. *American Psychologist* 49: 1004–1017.

Maizels, R.M. and Kurniawan-Atmadja, A. 2002. Variation and polymorphism in helminth parasites. *Parasitology* 125: S25–S37.

Manning, C.J., Wakeland, E.K. and Potts, W.K. 1992. Communal nesting patterns in mice implicate MHC genes in kin recognition. *Nature* 360: 581–583.

Marcogliese, D.J. 2001a. Pursuing parasites up the food chain: implications of food web structure and function on parasite communities in aquatic systems. *Acta Parasitologica* 46: 82–93.

Marcogliese, D.J. 2001b. Implications of climate change for parasitism of animals in the aquatic environment. *Canadian Journal of Zoology* 79: 1331–1352.

Marcogliese, D.J. 2002. Food webs and the transmission of parasites to marine fish. *Parasitology* 124: S83–S99.

Marcogliese, D.J. and Cone, D.K. 1991a. Do brook charr (*Salvelinus fontinalis*) from insular Newfoundland have different parasites than their mainland counterparts? *Canadian Journal of Zoology* 69: 809–811.

Marcogliese, D.J. and Cone, D.K. 1991b. Importance of lake characteristics in structuring parasite communities of salmonids from insular Newfoundland. *Canadian Journal of Zoology* 69: 2962–2967.

Marcogliese, D.J. and Cone, D.K. 1996. On the distribution and abundance of eel parasites in Nova Scotia: influence of pH. *Journal of Parasitology* 82: 389–399.

Marcogliese, D.J. and Cone, D.K. 1998. Comparison of richness and diversity of macroparasite communities among eels from Nova Scotia, the United Kingdom and Australia. *Parasitology* 116: 73–83.

Mardulyn, P. and Whitfield, J.B. 1999. Phylogenetic signal in the COI, 16S, and 28S genes for inferring relationships among genera of Microgastrinae (Hymenoptera; Braconidae): evidence of a high diversification rate in this group of parasitoids. *Molecular Phylogenetics and Evolution* 12: 282–294.

Margolis, L. and Arthur, J.R. 1979. *Synopsis of the Parasites of Fishes of Canada.* Bulletin of the Fisheries Research Board of Canada 199, Ottawa.

Martin, L.B., Scheuerlein, A. and Wikelski, M. 2003. Immune activity elevates energy expenditure of house sparrows: a link between direct and indirect costs? *Proceedings of the Royal Society of London* B 270: 153–158.

Martin, T.E., Møller, A.P., Merino, S. and Clobert, J. 2001. Does clutch size evolve in response to parasites and immunocompetence? *Proceedings of the National Academy of Sciences U.S.A.* 98: 2071–2076.

Marzluff, J.M. and Dial, K.P. 1991. Life history correlates of taxonomic diversity. *Ecology* 72: 428–439.

Maurer, B.A. 1999. *Untangling Ecological Complexity: The Macroscopic Perspective.* University of Chicago Press, Chicago.

Maurer, B.A., Brown, J.H. and Rusler, R.D. 1992. The micro and macro in body size evolution. *Evolution* 46: 939–953.

May, R.M. 1988. How many species are there on earth? *Science* 241: 1441–1449.

May, R.M. 1990. How many species? *Philosophical Transactions of the Royal Society of London* B 330: 293–304.

May, R.M. and Anderson, R.M. 1978. Regulation and stability of host-parasite population interactions. II. Destabilizing processes. *Journal of Animal Ecology* 47: 249–267.

Mayr, E. 1984. Evolution of fish species flocks. In: *Evolution of Fish Species Flocks* (Echelle, A.A. and Kornfield, I., editors). University of Maine at Orono Press, Orono, pp. 3–11.

McCoy, K.D. 2003. Sympatric speciation in parasites: what is sympatry? *Trends in Parasitology* 19: 400–404.

McGill, B.J. 2003. A test of the unified neutral theory of biodiversity. *Nature* 422: 881–885.

McGuire, K.L., Duncan, W.R. and Tucker, P.W. 1985. Syrian hamster DNA shows limited polymorphism at class I-like loci. *Immnunogenetics* 22: 257–258.

McNab, B.M. 1980. Food habits, energetics, and population biology of mammals. *American Naturalist* 116: 106–124.

McNab, B.K. 1992. A statistical analysis of mammalian rates of metabolism. *Functional Ecology* 6: 672–679.

Meagher, S. 1999. Genetic diversity and *Capillaria hepatica* (Nematoda) prevalence in Michigan deer mouse populations. *Evolution* 53: 1318–1324.

Mendonca, M.D. 2001. Galling insect diversity patterns: the resource synchronisation hypothesis. *Oikos* 95: 171–176.

Merrett, N.R. and Haedrich, R.L. 1997. *Deep-sea Demersal Fish and Fisheries*. Chapman and Hall, London.

Meyer, D. and Thomson, G. 2001. How selection shapes variation of the human major histocompatibility complex: a review. *Annals of Human Genetics* 65: 1–26.

Minchella, D.J. and Scott, M.E. 1991. Parasitism: a cryptic determinant of animal community structure. *Trends in Ecology and Evolution* 6: 250–254.

Mitter, C., Farrell, B. and Wiegemann, B. 1988. The phylogenetic study of adaptive zones: has phytophagy promoted insect diversification? *American Naturalist* 132: 107–128.

Møller, A.P. 1997. Parasitism and the evolution of host life history. In: *Host-Parasite Evolution. General Principles and Avian Models* (Clayton, D.H. and Moore, J., editors). Oxford University Press, Oxford, pp. 105–127.

Møller, A.P. 1998. Evidence of larger impact of parasites on hosts in the tropics: investment in immune function within and outside the tropics. *Oikos* 82: 265–270.

Møller, A.P. and Erritzøe, J. 1998. Host immune defense and migration in birds. *Evolutionary Ecology* 12: 945–953.

Møller, A.P., Sorci, G. and Erritzøe, J. 1998. Sexual dimorphism in immune defense. *American Naturalist* 152: 605–619.

Møller, A.P., Christe, P. and Lux, E. 1999. Parasitism, host immune function, and sexual selection. *Quarterly Review of Biology* 74: 3–30.

Møller, A.P., Merino, S., Brown, C.R. and Robertson, R. J. 2001. Immune defense and host sociality: A comparative study of swallows and martins. *American Naturalist* 158: 136–145.

Moore, J., Simberloff, D. and Freehling, M. 1988. Relationships between bobwhite quail social-group size and intestinal helminth parasitism. *American Naturalist* 131: 22–32.

Morand, S. 1993. Sexual transmission of a nematode: study of a model. *Oikos* 66: 48–54.

Morand, S. 1996a. Biodiversity of parasites in relation to their life cycles. In: *Aspects of the Genesis and Maintenance of Biological Diversity* (Hochberg, M.E., Clobert, J. and Barbault, R., editors). Oxford University Press, Oxford, pp. 243–260.

Morand, S. 1996b. Life-history traits in parasitic nematodes: a comparative approach for the search of invariants. *Functional Ecology* 10: 210–218.

Morand, S. 2000. Wormy world: comparative tests of theoretical hypotheses on parasite species richness. In: *Evolutionary Biology of Host-Parasite Relationships:*

Theory Meets Reality (Poulin, R., Morand, S. and Skorping, A., editors). Elsevier Science, Amsterdam, pp. 63–79.

Morand, S. 2003. Parasites and the evolution of host life history traits. In: *Proceedings of the 18th International Congress of Zoology* (Legakis, A., Sfenthourakis, S., Polymeni R. and Thessalou-Legaki, M., editors). Pensoft, Sofia, pp. 213–218.

Morand, S., Cribb, T.H., Kulbicki, M., Chauvet, C., Dufour, V., Faliex, E., Galzin, R., Lo, C., Lo-Yat, A., Pichelin, S.P., Rigby, M.C. and Sasal, P. 2000. Determinants of endoparasite species richness of New Caledonian Chaetodontidae. *Parasitology* 121: 65–73.

Morand, S. and Gonzales, E.A. 1997. Is parasitism a missing ingredient in model ecosystems? *Ecological Modelling* 95: 61–74.

Morand, S. and Guégan, J.-F. 2000. Distribution and abundance of parasite nematodes: ecological specialisation, phylogenetic constraint or simply epidemiology? *Oikos* 88: 563–573.

Morand, S. and Harvey, P.H. 2000. Mammalian metabolism, longevity and parasite species richness. *Proceedings of the Royal Society of London* B 267: 1999–2003.

Morand, S. and Poulin, R. 1998. Density, body mass and parasite species richness of terrestrial mammals. *Evolutionary Ecology* 12: 717–727.

Morand, S. and Poulin, R. 2000. Nematode parasite species richness and the evolution of spleen size in birds. *Canadian Journal of Zoology* 78: 1356–1360.

Morand, S. and Poulin, R. 2002. Body size-density relationships and species diversity in parasitic nematodes: patterns and likely processes. *Evolutionary Ecology Research* 4: 951–961.

Morand, S., Poulin, R., Rohde, K. and Hayward, C. 1999. Aggregation and species coexistence of ectoparasites of marine fishes. *International Journal for Parasitology* 29: 663–672.

Morand, S., Rohde, K. and Hayward, C. 2002. Order in ectoparasite communities of marine fish is explained by epidemiological processes. *Parasitology* 124: S57–S63.

Moreira, D. and López-Garcia, P. 2003. Are hydrothermal vents oases for parasitic protists? *Trends in Parasitology* 19: 556–558.

Moreno, J., Potti, J., Yorio, P. and Garcia, P. 2001. Sex differences in cell-mediated immunity in the Magellanic penguin *Spheniscus magelanicus*. *Annals Zoologici Fennica*. 38: 111–116.

Moreno, J., Sanz, J.J. and Arriero, E. 1999. Reproductive effort and T-lymphocyte cell-mediated immunocompetence in female pied flycatchers *Ficedula hypoleuca*. *Proceedings of the Royal Society of London* B 266: 1105–1109.

Moritz, C., McCallum, H., Donnellan, S. and Roberts, J.D. 1991. Parasite loads in parthenogenetic and sexual lizards (*Heteronotia binoei*): support for the Red Queen hypothesis. *Proceedings of the Royal Society of London* B 244: 145–149.

Morse, D.R, Lawton, J.H., Dodson, M.M. and Williamson, M.H. 1985. Fractal dimensions of vegetation and the distribution of arthropod body lengths. *Nature* 314: 731–733.

Mouillot, D., George-Nascimento, M. and Poulin, R. 2003. How parasites divide resources: a test of the niche apportionment hypothesis. *Journal of Animal Ecology* 72: 757–764.

Mouillot, D. and Poulin, R. 2004. Taxonomic partitioning shedding light on the diversification of parasite communities. *Oikos* 104: 205–207.

Mouritsen, K.N. and Poulin, R. 2002. Parasitism, community structure and biodiversity in intertidal ecosystems. *Parasitology* 124: S101–S117.

Nascetti, G., Cianchi, R., Mattiucci, S., D'Amelio, S., Orecchia, P., Paggi, L., Brattey, J., Berland, B., Smith, J.W. and Bullini, L. 1993. Three sibling species within *Contracaecum osculatum* (Nematoda, Ascaridida, Ascaridoidea) from the Atlantic Arctic-Boreal region: reproductive isolation and host preferences. *International Journal for Parasitology* 23: 105–120.

Nee, S., Barraclough, T.G. and Harvey, P.H. 1996. Temporal changes in biodiversity: detecting patterns and identifying causes. In: *Biodiversity: A Biology of Numbers and Difference* (Gaston, K.J., editor). Blackwell Science, Oxford, pp. 230–252.

Nee, S. and Stone, G. 2003. The end of the beginning for the neutral theory. *Trends in Ecology and Evolution* 18: 433–434.

Nei, M. and Hugues, A.L. 1991. Polymorphism and evolution of the major histocompatibility complex loci in mammals. In: *Evolution at the Molecular Level* (Selander, R.K., Clark, A.G. and Whittam, T.S., editors). Sinauer, Sunderland, MA, pp. 222–247.

Nekola, J.C. and White, P.S. 1999. The distance decay of similarity in biogeography and ecology. *Journal of Biogeography* 26: 867–878.

Noble, E.R. 1973. Parasites and fishes in a deep-sea environment. *Advances in Marine Biology* 11: 121–195.

Noble, E.R., Noble, G.A., Schad, G.A. and MacInnes, A.J. 1989. *Parasitology: The Biology of Animal Parasites*, sixth edition. Lea and Febiger, Philadelphia.

Nordling, D., Andersson, M., Zohari, S. and Gustafsson, L. 1998. Reproductive effort reduces specific immune response and parasite resistance. *Proceedings of the Royal Society of London* B 265: 1291–1298.

Novotny, V., Basset, Y., Miller, S.E., Weiblen, G.D., Bremer, B., Cizek, L. and Drozd, P. 2002. Low host specificity of herbivorous insects in a tropical forest. *Nature* 416: 841–844.

Nunn, C.L., Altizer, S., Jones, K.E. and Sechrest, W. 2003a. Comparative tests of parasite species richness in primates. *American Naturalist* 162: 597–614.

Nunn, C.L., Gittleman, J.L. and Antonovics, J. 2003b. A comparative study of white blood cell counts and disease risk in carnivores. *Proceedings of the Royal Society of London* B 270: 347–356.

Oakley, T.H. and Cunningham, C.W. 2000. Independent contrasts succeed where ancestor reconstruction fails in a known bacteriophage phylogeny. *Evolution* 54: 397–405.

O'Brien, C.W. and Wibmer, G.J. 1979. The use of trend curves of rates of species descriptions: examples from the Curculionidae (Coleoptera). *The Coleopterist's Bulletin* 33: 151–166.

Ødegaard, F. 2000. How many species of arthropods? Erwin's estimate revisited. *Biological Journal of the Linnean Society* 71: 583–597.

Okamura, B. and Canning, E.U. 2003. Orphan worms and homeless parasites enhance bilaterian diversity. *Trends in Ecology and Evolution* 18: 633–639.

Orme, C.D.L., Isaac, N.J.B. and Purvis, A. 2002a. Are most species small? Not within species-level phylogenies. *Proceedings of the Royal Society of London* B 269: 1279–1287.

Orme, C.D.L., Quicke, D.L.J., Cook, J.M. and Purvis, A. 2002b. Body size does not predict species richness among the metazoan phyla. *Journal of Evolutionary Biology* 15: 235–247.

Osler, G.H.R. and Beattie, A.J. 2001. Contribution of oribatid and mesostigmatid soil mites in ecologically based estimates of global species richness. *Austral Ecology* 26: 70–79.

Ots, I., Erimov, A.B., Ivankina, E.V., Ilyina, T.A. and Horak, P. 2001. Immune challenge affects basal metabolic activity in wintering great tits. *Proceedings of the Royal Society of London* B 268: 1175–1181.

Owens, I.P.F. and Wilson, K. 1999. Immunocompetence: a neglected life history trait or conspicuous red herring? *Trends in Ecology and Evolution* 14: 170–172.

Page, R.D.M. 1990. Temporal congruence and cladistic analysis of biogeography and cospeciation. *Systematic Zoology* 39: 205–226.

Page, R.D.M. 1993. Parasites, phylogeny and cospeciation. *International Journal for Parasitology* 23: 499–506.

Page, R.D.M. 1994. Parallel phylogenies: reconstructing the history of host-parasite assemblages. *Cladistics* 10: 155–173.

Page, R.D.M. 2003. *Tangled Trees: Phylogeny, Cospeciation, and Coevolution.* University of Chicago Press, Chicago.

Palmer, M.W. 1990. The estimation of species richness by extrapolation. *Ecology* 71: 1195–1198.

Pariselle, A. 1996. *Diversité, spéciation et évolution des Monogènes branchiaux de Cichlidae en Afrique de l'Ouest.* PhD Thesis, Université de Perpignan, France.

Paterson, A.M. and Banks, J. 2001. Analytical approaches to measuring cospeciation of host and parasites: through a glass, darkly. *International Journal for Parasitology* 31: 1012–1022.

Paterson, A.M. and Gray, R.D. 1997. Host-parasite cospeciation, host switching, and missing the boat. In: *Host-Parasite Evolution: General Principles and Avian Models* (Clayton, D.H. and Moore, J., editors). Oxford University Press, Oxford, pp. 236–250.

Paterson, A.M., Gray, R.D. and Wallis, G.P. 1993. Parasites, petrels and penguins: does louse presence reflect seabird phylogeny? *International Journal for Parasitology* 23: 515–526.

Paterson, A.M., Palma, R.L. and Gray, R.D. 1999. How frequently do avian lice miss the boat? Implications for coevolutionary studies. *Systematic Biology* 48: 214–223.

Paterson, A.M., Palma, R.L. and Gray, R.D. 2003. Drowning on arrival, missing the boat, and x-events: how likely are sorting events? In: *Tangled Trees: Phylogeny, Cospeciation, and Coevolution* (Page, R.D.M., editor). University of Chicago Press, Chicago, pp. 287–309.

Paterson, A.M. and Poulin, R. 1999. Have chondracanthid copepods co-speciated with their teleost hosts? *Systematic Parasitology* 44: 79–85.

Paterson, S., Wilson, K. and Pemberton, J.M. 1998. Major histocompatibility complex variation associated with juvenile survival and parasite resistance in a large unmanaged ungulate population (*Ovis aries* L.). *Proceedings of the National Academy of Sciences U.S.A.* 95: 3714–3719.

Patrick, M.J. 1991. Distribution of enteric helminths in *Glaucomys volans* L. (Sciuridae): a test for competition. *Ecology* 72: 755–758.

Patterson, B.D. and Atmar, W. 1986. Nested subsets and the structure of insular mammalian faunas and archipelagos. *Biological Journal of the Linnean Society* 28: 65–82.

Penn, D.J. and Potts, W.K. 1998. Chemical signals and parasite-mediated sexual selection. *Trends in Ecology and Evolution* 13: 391–396.

Pérez Ponce de León, G. 1995. Host-induced morphological variability in adult *Posthodiplostomum minimum* (Digenea: Neodiplostomidae). *Journal of Parasitology* 81: 818–820.

Pérez Ponce de León, G. 2001. The diversity of digeneans (Platyhelminthes: Cercomeria: Trematoda) in vertebrates in Mexico. *Comparative Parasitology* 68: 1–8.

Pérez Ponce de León, G., Garcia-Prieto, L. and Razo-Mendivil, U. 2002. Species richness of helminth parasites in Mexican amphibians and reptiles. *Diversity and Distributions* 8: 211–218.

Petchey, O.L. and Gaston, K.J. 2002. Functional diversity (FD), species richness and community composition. *Ecology Letters* 5: 402–411.

Peters, R.H. 1983. *The Ecological Implications of Body Size*. Cambridge University Press, Cambridge.

Pfau, R.S., van den Bussche, R.A. and McBee, K. 2001. Population genetics of the hispid cotton rat (*Sigmodon hispidus*): patterns of genetic diversity at the major histocompatibility complex. *Molecular Ecology* 10: 1939–1945.

Pianka, E.R. 1966. Latitudinal gradients in species diversity: a review of the concepts. *American Naturalist* 100: 33–46.

Pisanu, B., Chapuis, J.L. and Durette-Desset, M.C. 2001. Helminths from introduced small mammals on Kerguelen, Crozet, and Amsterdam Islands (Southern Indian Ocean). *Journal of Parasitology* 87: 1205–1208.

Platt, T.R. and Jensen, R.J. 2002. *Aptorchis aequalis* Nicoll, 1914 (Digenea: Plagiorchiidae) is a senior synonym of *Dingularis anfracticirrus* Jue Sue and Platt, 1999 (Digenea: Plagiorchiidae). *Systematic Parasitology* 52: 183–191.

Poiani, A. 1992. Ectoparasitism as a possible cost of social life: a comparative analysis using Australian passerines (Passeriformes). *Oecologia* 92: 429–441.

Poinar, G. Jr. 1999. *Paleochordodes protus* n.g., n.sp. (Nematomorpha, Chordodidae), parasites of a fossil cockroach, with a critical examination of other fossil hairworms and helminths of extant cockroaches (Insecta: Blattaria). *Invertebrate Biology* 118: 109–115.

Poinar, G. Jr., Krantz, G.W., Boucot, A.J. and Pike, T.M. 1997. A unique Mesozoic parasitic association. *Naturwissenschaften* 84: 321–322.

Potts, W.K., Manning, C.J. and Wakeland, E.K. 1991. Mating patterns in semi-natural populations of mice influenced by MHC genotype. *Nature* 352: 619–621.

Potts, W.K., Manning, C.J. and Wakeland, E.K. 1994. The role of infectious disease, inbreeding, and mating preferences in maintaining MHC diversity: an experimental test. *Philosophical Transactions of the Royal Society of London* B 346: 369–378.

Poulin, R. 1991. Group-living and the richness of the parasite fauna in Canadian freshwater fishes. *Oecologia* 86: 390–394.

Poulin, R. 1992a. Determinants of host specificity in parasites of freshwater fishes. *International Journal for Parasitology* 22: 753–758.

Poulin, R. 1992b. Toxic pollution and parasitism in freshwater fish. *Parasitology Today* 8: 58–61.

Poulin, R. 1995a. Clutch size and egg size in free-living and parasitic copepods: a comparative analysis. *Evolution* 49: 325–336.

Poulin, R. 1995b. Evolutionary influences on body size in free-living and parasitic isopods. *Biological Journal of the Linnean Society* 54: 231–244.

Poulin, R. 1995c. Phylogeny, ecology, and the richness of parasite communities in vertebrates. *Ecological Monographs* 65: 283–302.

Poulin, R. 1996a. Richness, nestedness, and randomness in parasite infracommunity structure. *Oecologia* 105: 545–551.

Poulin, R. 1996b. How many parasite species are there: are we close to answers? *International Journal for Parasitology* 26: 1127–1129.

Poulin, R. 1996c. The evolution of life history strategies in parasitic animals. *Advances in Parasitology* 37: 107–134.

Poulin, R. 1996d. The evolution of body size in the Monogenea: the role of host size and latitude. *Canadian Journal of Zoology* 74: 726–732.

Poulin, R. 1997a. Parasite faunas of freshwater fish: the relationship between richness and the specificity of parasites. *International Journal for Parasitology* 27: 1091–1098.

Poulin, R. 1997b. Species richness of parasite assemblages: evolution and patterns. *Annual Review of Ecology and Systematics* 28: 341–358.

Poulin, R. 1998a. *Evolutionary Ecology of Parasites: From Individuals to Communities.* Chapman and Hall, London.

Poulin, R. 1998b. Comparison of three estimators of species richness in parasite component communities. *Journal of Parasitology* 84: 485–490.

Poulin, R. 1998c. Large-scale patterns of host use by parasites of freshwater fishes. *Ecology Letters* 1: 118–128.

Poulin, R. 1999a. Speciation and diversification of parasite lineages: an analysis of congeneric parasite species in vertebrates. *Evolutionary Ecology* 13: 455–467.

Poulin, R. 1999b. The intra- and interspecific relationships between abundance and distribution in helminth parasites of birds. *Journal of Animal Ecology* 68: 719–725.

Poulin, R. 2001. Another look at the richness of helminth communities in tropical freshwater fish. *Journal of Biogeography* 28: 737–743.

Poulin, R. 2002. The evolution of monogenean diversity. *International Journal for Parasitology* 32: 245–254.

Poulin, R. 2003. The decay of similarity with geographical distance in parasite communities of vertebrate hosts. *Journal of Biogeography* 30: 1609–1615.

Poulin, R. 2004a. Parasite species richness in New Zealand fishes: a grossly underestimated component of biodiversity? *Diversity and Distributions* 10: 31–37.

Poulin, R. 2004b. Parasites and the neutral theory of biodiversity. *Ecography* 27: 119–123.

Poulin, R. and Cribb, T.H. 2002. Trematode life cycles: short is sweet? *Trends in Parasitology* 18: 176–183.

Poulin, R. and Guégan, J.-F. 2000. Nestedness, anti-nestedness, and the relationship between prevalence and intensity in ectoparasite assemblages of marine fish: a spatial model of species coexistence. *International Journal for Parasitology* 30: 1147–1152.

Poulin, R. and Hamilton, W.J. 1995. Ecological determinants of body size and clutch size in amphipods: a comparative approach. *Functional Ecology* 9: 364–370.

Poulin, R., Marshall, L.J. and Spencer, H.G. 2000. Metazoan parasite species richness and genetic variation among freshwater fish species: cause or consequence? *International Journal for Parasitology* 30: 697–703.

Poulin, R. and Morand, S. 1997. Parasite body size distributions: interpreting patterns of skewness. *International Journal for Parasitology* 27: 959–964.

Poulin, R. and Morand, S. 1999. Geographic distances and the similarity among parasite communities of conspecific host populations. *Parasitology* 119: 369–374.

Poulin, R. and Morand, S. 2000. The diversity of parasites. *Quarterly Review of Biology* 75: 277–293.

Poulin, R. and Mouillot, D. 2003. Host introductions and the geography of parasite taxonomic diversity. *Journal of Biogeography* 30: 837–845.

Poulin, R. and Mouillot, D. 2004. The evolution of taxonomic diversity in helminth assemblages of mammalian hosts. *Evolutionary Ecology* 18: 231–247.

Poulin, R., Mouillot, D. and George-Nascimento, M. 2003. The relationship between species richness and productivity in metazoan parasite communities. *Oecologia* 137: 277–285.

Poulin, R. and Mouritsen, K.N. 2003. Large-scale determinants of trematode infections in intertidal gastropods. *Marine Ecology Progress Series* 254: 187–198.

Poulin, R. and Rohde, K. 1997. Comparing the richness of metazoan ectoparasite communities of marine fishes: controlling for host phylogeny. *Oecologia* 110: 278–283.

Poulin, R. and Thomas, F. 1999. Phenotypic variability induced by parasites: extent and evolutionary implications. *Parasitology Today* 15: 28–32.

Price, P.W. 1980. *Evolutionary Biology of Parasites*. Princeton University Press, Princeton.

Price, P.W. 2002. Species interactions and the evolution of biodiversity. In: *Plant-Animal Interactions: An Evolutionary Approach* (Herrera, C.M. and Pellmyr, O., editors). Blackwell Science, Oxford, pp. 3–25.

Price, P.W. and Clancy, K.M. 1983. Patterns in number of helminth parasite species in freshwater fishes. *Journal of Parasitology* 69: 449–454.

Price, P.W., Fernandes, G.W., Lara, A.C.F., Brawn, J., Barrios, H., Wright, M.G., Ribeiro, S.P. and Rothcliff, N. 1998. Global patterns in local number of insect galling species. *Journal of Biogeography* 25: 581–591.

Purvis, A. and Hector, A. 2000. Getting the measure of biodiversity. *Nature* 405: 212–219.

Raberg, L., Grahm, M., Hasselquist, D. and Svensson, E. 1998. On the adaptive significance of stress-induced immunosuppression. *Proceedings of the Royal Society of London* B 265: 1637–1641.

Raberg, L., Vestberg, M., Hasselquist, D., Holmdahl, R., Svensson, E. and Nilsson, J.-A. 2002. Basal metabolic rate and the evolution of the adaptive immune system. *Proceedings of the Royal Society of London* B 269: 817–821.

Raibaut, A., Combes, C. and Benoit, F. 1998. Analysis of the parasitic copepod species richness among Mediterranean fish. *Journal of Marine Systems* 15: 185–206.

Ranta, E. 1992. Gregariousness versus solitude: another look at parasite faunal richness in Canadian freshwater fishes. *Oecologia* 89: 150–152.

Reed, R.N. and Boback, S.M. 2002. Does body size predict dates of species description among North American and Australian reptiles and amphibians? *Global Ecology and Biogeography* 11: 41–47.

Renaud, F. and Gabrion, C. 1988. Speciation in Cestoda: evidence of two sibling species in the complex *Bothrimonus nylandicus* (Scheider, 1902) (Cestoda, Pseudophyllidea). *Parasitology* 97: 1–9.

Reversat, J., Maillard, C. and Silan, P. 1991. Polymorphisme phénotypique et enzymatique: intérêt et limites dans la description d'espèces d'*Helicometra* (Trematoda: Opecoelidae), mésoparasites de téléostéens marins. *Systematic Parasitology* 19: 147–158.

Richman, A. 2000. Evolution of balanced genetic polymorphism. *Molecular Ecology* 9: 1953–1963.

Ricklefs, R.E. 1992. Embryonic development period and the prevalence of avian blood parasites. *Proceedings of the National Academy of Sciences U.S.A.* 89: 4722–4725.

Ricklefs, R.E. and Fallon, S.M. 2002. Diversification and host switching in avian malaria parasites. *Proceedings of the Royal Society of London* B 269: 885–892.

Ricklefs, R.E. and Schluter, D. 1993. *Species Diversity in Ecological Communities: Historical and Geographical Perspectives.* University of Chicago Press, Chicago.

Rigby, M.C., Holmes, J.C., Cribb, T.H. and Morand, S. 1997. Patterns in the gastrointestinal helminths of a coral reef-associated fish, *Epinephelus merra* (Serranidae), from French Polynesia and the South Pacific. *Canadian Journal of Zoology* 75: 1818–1827.

Rigby, M.C. and Moret, Y. 2000. Life-history trade-offs with immune defenses. In: *Evolutionary Biology of Host-Parasite Relationships: Theory Meets Reality* (Poulin, R., Morand, S. and Skorping, A., editors). Elsevier, Amsterdam, pp. 129–142.

Roberts, L.S. and Janovy, J. Jr. 1996. *Foundations of Parasitology,* fifth edition. W.C. Brown Publishers, Dubuque, Iowa.

Roberts, M.G. and Dobson, A.P. 1995. The population dynamics of communities of parasitic helminths. *Mathematical Biosciences* 126: 191–214.

Roberts, M.G., Dobson, A.P., Arneberg, P., de Leo, G.A., Krecek, R.C., Manfredi, M.T., Lanfranchi, P. and Zaffaroni, E. 2002. Parasite community ecology and biodiversity. In: *The Ecology of Wildlife Diseases* (Hudson, P.J., Rizzoli, A., Grenfell, B.T., Heesterbeek, H. and Dobson, A.P., editors). Oxford University Press, Oxford, pp. 63–82.

Roff, D. 1992. *The Evolution of Life Histories: Theory and Analysis.* Chapman and Hall, New York.

Rohde, K. 1978. Latitudinal differences in host specificity of marine Monogenea and Digenea. *Marine Biology* 47: 125–134.

Rohde, K. 1980. Diversity gradients of marine Monogenea in the Atlantic and Pacific Oceans. *Experientia* 36: 1368–1369.

Rohde, K. 1986. Differences in species diversity of Monogenea between the Pacific and Atlantic oceans. *Hydrobiologia* 137: 21–28.

Rohde, K. 1992. Latitudinal gradients in species diversity: the search for the primary cause. *Oikos* 65: 514–527.

Rohde, K. 1993. *Ecology of Marine Parasites,* second edition. CAB International, Wallingford, UK.

Rohde, K. 1996. Robust phylogenies and adaptive radiations: a critical examination of methods used to identify key innovations. *American Naturalist* 148: 481–500.

Rohde, K. 1997. The larger area of the tropics does not explain latitudinal gradients in species diversity. *Oikos* 79: 169–172.

Rohde, K. 1998a. Is there a fixed number of niches for endoparasites of fishes? *International Journal for Parasitology* 28: 1861–1865.

Rohde, K. 1998b. Latitudinal gradients in species diversity: area matters, but how much? *Oikos* 82: 184–190.

Rohde, K. 1999. Latitudinal gradients in species diversity and Rapoport's rule revisited: a review of recent work and what can parasites teach us about the causes of the gradients? *Ecography* 22: 593–613.

Rohde, K. 2002. Ecology and biogeography of marine parasites. *Advances in Marine Biology* 43: 1–86.

Rohde, K. and Hayward, C.J. 2000. Oceanic barriers as indicated by scombrid fiches and their parasites. *International Journal for Parasitology* 30: 579–583.

Rohde, K., Hayward, C.J. and Heap. M. 1995. Aspects of the ecology of metazoan ectoparasites of marine fishes. *International Journal for Parasitology* 25: 945–970.

Rohde, K. and Heap, M. 1998. Latitudinal differences in species and community richness and in community structure of metazoan endo- and ectoparasites of marine teleost fish. *International Journal for Parasitology* 28: 461–474.

Rosenzweig, M.L. 1995. *Species Diversity in Space and Time*. Cambridge University Press, Cambridge.

Rosenzweig, M.L. and Sandlin, E.A. 1997. Species diversity and latitudes: listening to area's signal. *Oikos* 80: 172–176.

Rózsa, L. 1992. Endangered parasite species. *International Journal for Parasitology* 22: 265–266.

Sasal, P. and Morand, S. 1998. Comparative analysis: a tool for studying monogenean ecology and evolution. *International Journal for Parasitology* 28: 1637–1644.

Sasal, P., Morand, S. and Guégan, J.-F. 1997. Determinants of parasite species richness in Mediterranean marine fish. *Marine Ecology Progress Series* 149: 61–71.

Schad, G.A. 1963. Niche diversification in a parasite species flock. *Nature* 198: 404–406.

Schluter, D. 2000. *The Ecology of Adaptive Radiation*. Oxford University Press, Oxford.

Schmid-Hempel, P. 1998. *Parasites in Social Insects*. Princeton University Press, Princeton.

Schmid-Hempel, P. 2001. On the evolutionary ecology of host-parasite interactions: addressing the question with regard to bumblebees and their parasites. *Naturwissenschaften* 88: 147–158.

Schmid-Hempel, P. 2003. Variation in immune defense as a question of evolutionary ecology. *Proceedings of the Royal Society of London* B 270: 357–366.

Schmid-Hempel, P. and Crozier, R.H. 1999. Polyandry versus polygyny versus parasites. *Philosophical Transactions of the Royal Society of London* B 354: 507–515.

Schmidt-Nielsen, K. 1984. *Scaling: Why is Animal Size so Important?* Cambridge University Press, Cambridge.

Schmidt-Rhaesa, A. 1997. *Nematomorpha*. Freshwater Fauna of Central Europe, Volume 4/4. Gustav Fischer Verlag, Stuttgart.

Scott, M.E. 1987. Regulation of mouse colony abundance by *Heligmosomoides polygyrus*. *Parasitology* 95: 111–124.

Seddom, J.M. and Baverstock, P.R. 1999. Variation on islands: major histocompatibility complex (MHC) polymorphism in populations of Australian bush rat. *Molecular Ecology* 8: 2071–2079.

Shaw, D.J. and Dobson, A.P. 1995. Patterns of macroparasite abundance and aggregation in wildlife populations: a quantitative review. *Parasitology* 111: S111–S133.

Shaw, D.J. Grenfell, B.T. and Dobson, A.P. 1998. Patterns of macroparasite aggregation in wildlife host populations. *Parasitology* 117: 597–610.

Shorrocks, B., Marsters, J., Ward, I. and Evennett, P.J. 1991. The fractal dimension of lichens and the distribution of arthropod body lengths. *Functional Ecology* 5: 457–460.

Shutler, D., Alisauskas, R.T. and McLaughlin, J.D. 1999. Mass dynamics of the spleen and other organs in geese: measures of immune relationships to helminths? *Canadian Journal of Zoology* 77: 351–359.

Siddall, M.E. and Burreson, E.M. 1998. Phylogeny of leeches (Hirudinea) based on mitochondrial cytochrome c oxydase subunit I. *Molecular Phylogenetics and Evolution* 9: 156–162.

Siddall, M.E. and Perkins, S.L. 2003. Brooks Parsimony Analysis: a valiant failure. *Cladistics* 19: 554–564.

Silver, B.B., Dick, T.A. and Welch, H.E. 1980. Concurrent infections of *Hymenolepis diminuta* and *Trichinella spiralis* in the rat intestine. *Journal of Parasitology* 66: 786–791.

Simkova, A., Kadlec, D., Gelnar, M. and Morand, S. 2002. Abundance-prevalence relationship of gill congeneric ectoparasites: testing for the core-satellite hypothesis and ecological specialisation. *Parasitology Research* 88: 682–686.

Simkova, A., Morand, S., Jobet, E., Gelnar, M. and Verneau, O. 2004. Molecular phylogeny of congeneric monogenean parasites (*Dactylogyrus*): a case of intrahost speciation. *Evolution* 58: 1001–1018.

Simkova, A., Morand, S., Matejusova, I., Jurajda, P.V. and Gelnar, M. 2001. Local and regional influences on patterns of parasite species richness of central European fishes. *Biodiversity and Conservation* 10: 511–525.

Sinclair, J.A. and Lochmiller, R.L. 2000. The winter immunoenhancement hypothesis: associations among immunity, density, and survival in prairie vole (*Microtus ochrogaster*) populations. *Canadian Journal of Zoology* 78: 254–264.

Sisk, T.D., Launer, A.E., Switky, K.R. and Ehrlich, P.R. 1994. Identifying extinction threats. *BioScience* 44: 592–604.

Skarstein, F. and Folstad, I. 1996. Sexual dichromatism and immunocompetence handicap: an observational approach using Arctic charr. *Oikos* 76: 359–367.

Skerikova, A., Hypsa, V. and Scholz, T. 2001. Phylogenetic analysis of European species of *Proteocephalus* (Cestoda: Proteocephalidea): compatibility of molecular and morphological data, and parasite-host coevolution. *International Journal for Parasitology* 31: 1121–1128.

Skorping, A., Read, A.F. and Keymer, A.E. 1991. Life history covariation in intestinal nematodes of mammals. *Oikos* 60: 365–372.

Skrjabin, K.I. 1953–1965. *Principles of Nematology.* Academy NAUK SSSR, Moscow (in Russian).

Slade, R. 1992. Limited MHC polymorphism in the southern elephant seal: implications for MHC evolution and marine mammal population biology. *Proceedings of the Royal Society of London* B 249: 163–171.

Slade, R. and McCallum, H.I. 1992. Overdominant vs frequency-dependent selection at MHC loci. *Genetics* 132: 861–862.

Slater, C.H., Fitzpatrick, M.S. and Schreck, C.B. 1995. Androgens and immunocompetence in salmonids: specific binding in and reduced immunocompetence of salmonid lymphocytes exposed to natural and synthetic androgens. *Aquaculture* 136: 363–370.

Slowinski, J.B. and Guyer, C. 1989. Testing the stochasticity of patterns of organismal diversity: an improved null model. *American Naturalist* 134: 907–921.

Slowinski, J.B. and Guyer, C. 1993. Testing whether certain traits have caused amplified diversification: an improved method based on a model of random speciation and extinction. *American Naturalist* 142: 1019–1024.

Smith, E.P. and van Belle, G. 1984. Nonparametric estimation of species richness. *Biometrics* 40: 119–129.

Smith, F.D.M., May, R.M., Pellew, R., Johnson, T.H. and Walter, K.R. 1993a. How much do we know about the current extinction rate? *Trends in Ecology and Evolution* 8: 375–378.

Smith, F.D.M., May, R.M., Pellew, R., Johnson, T.H. and Walter, K.R. 1993b. Estimating extinction rates. *Nature* 364: 494–496.

Smits, J.E., Bortolotti, G.R. and Tella, J.L. 1999. Simplifying the phytohaemagglutinin skin-testing technique in studies of avian immunocompetence. *Functional Ecology* 13: 567–572.

Smits, J.E., Bortolotti, G.R. and Tella, J.L. 2001. Measurement repeatability and the use of controls in PHA assays: a reply to Siva-Jothy and Ryder. *Functional Ecology* 15: 814–817.

Soberón, J. and Llorente, J. 1993. The use of species accumulation functions for the prediction of species richness. *Conservation Biology* 7: 480–488.

Soler, J.J., de Neve, L., Pérez-Contreras, T., Soler, M. and Sorci, G. 2003. Trade-off between immunocompetence and growth in magpies: an experimental study. *Proceedings of the Royal Society of London* B 270: 241–248.

Solow, A.R., Mound, L.A. and Gaston, K.J. 1995. Estimating the rate of synonymy. *Systematic Biology* 44: 93–96.

Sommer, S., Schwab, D. and Ganzhorn, J.U. 2002. MHC diversity of endemic Malagasy rodents in relation to geographic range and social system. *Behavioral Ecology and Sociobiology* 51: 214–221.

Sorci, G., Skarstein, F., Morand, S. and Hugot, J.-P. 2003. Correlated evolution between host immunity and parasite life histories in primates and oxyurid parasites. *Proceedings of the Royal Society of London* B 270: 2481–2484.

Sousa, W.P. 1993. Interspecific antagonism and species coexistence in a diverse guild of larval trematode parasites. *Ecological Monographs* 63: 103–128.

Southwood, T.R.E. 1987. Species-time relationships in human parasites. *Evolutionary Ecology* 1: 245–246.

Southwood, T.R.E. and Kennedy, C.E.J. 1983. Trees as islands. *Oikos* 41: 359–371.

Sprent, J.F.A. 1992. Parasites lost. *International Journal for Parasitology* 22: 139–151.

Srivastava, D.S. 1999. Using local-regional richness plots to test for species saturation: pitfalls and potentials. *Journal of Animal Ecology* 68: 1–16.

Stankiewicz, M., Cowan, P.E. and Heath, D.D. 1997a. Endoparasites of brushtail possums (*Trichosurus vulpecula*) from the South Island, New Zealand. *New Zealand Veterinary Journal* 45: 257–260.

Stankiewicz, M., Heath, D.D. and Cowan, P.E. 1997b. Internal parasites of possums (*Trichosurus vulpecula*) from Kawau Island, Chatham Island and Stewart Island. *New Zealand Veterinary Journal* 45: 247–250.

Stanko, M., Miklisova, D., Goüy de Bellocq, J. and Morand, S. 2002. Mammal density and patterns of ectoparasite species richness and abundance. *Oecologia* 131: 289–295.

Stearns, S.C. 1992. *The Evolution of Life Histories.* Oxford University Press, Oxford.

Stearns, S.C. and Koella, J. 1986. The evolution of phenotypic plasticity in life-history traits: predictions for norms of reaction for age- and size-at-maturity. *Evolution* 40: 893–913.

Stevens, G.C. 1989. The latitudinal gradient in geographical range: how so many species co-exist in the tropics. *American Naturalist* 133: 240–256.

Stork, N.E. 1988. Insect diversity: facts, fiction and speculation. *Biological Journal of the Linnean Society* 35: 321–337.

Stork, N.E. and Lyal, C.H.C. 1993. Extinction or 'co-extinction' rates? *Nature* 366: 307.

Strong, D.R., Lawton, J.H. and Southwood, R. 1984. *Insects on Plants: Community Patterns and Mechanisms.* Blackwell Science, Oxford.

Summers, K., McKeon, S., Sellars, J., Keusenkothen, M., Morris, J., Gloeckner, D., Pressley, C., Price, B. and Snow, H. 2003. Parasitic exploitation as an engine of diversity. *Biological Reviews* 78: 639–675.

Sutherland, W.J. 1998. *Conservation Science and Action.* Blackwell Science. Oxford.

Svensson, E., Raberg, L., Koch, C. and Hasselquist, D. 1998. Energetic costs of immune responses: implications for resource allocation and adaptive immunosuppression. *Functional Ecology* 12: 919–919.

Tauber, C.A. and Tauber, M.J. 1989. Sympatric speciation in insects: perception and perspective. In: *Speciation and its Consequences* (Otte, D. and Endler, J.A., editors). Sinauer Press, Sunderland, MA, pp. 307–344.

Taylor, J. and Purvis, A. 2003. Have mammals and their chewing lice diversified in parallel? In: *Tangled Trees: Phylogeny, Cospeciation, and Coevolution* (Page, R.D.M., editor). University of Chicago Press, Chicago, pp. 240–261.

Taylor, L.H., Latham, S.M. and Woolhouse, M.E.J. 2001. Risk factors for human disease emergence. *Philosophical Transactions of the Royal Society of London* B 356: 983–989.

Taylor, L.R. 1961. Aggregation, variance and the mean. *Nature* 189: 732–735.

Taylor, L.R., Woiwod, I.P. and Perry, J.N. 1978. The density-dependence of spatial behaviour and the rarity of randomness. *Journal of Animal Ecology* 47: 383–406.

Taylor, L.R., Compagno, L.J.V. and Struhsaker, P.J. 1983. Megamouth: a new species, genus, and family of lamnoid shark (*Megachasma pelagios,* family Megachasmidae) from Hawaiian Islands. *Proceedings of the California Academy of Sciences* 43A: 87–110.

Tella, J.L., Bortolotti, G.B., Dawson, R.D. and Forero, M.G. 2000. The T-cell mediated immune response and return rate of fledgling American kestrels are positively correlated with parental clutch size. *Proceedings of the Royal Society of London* B 267: 891–895.

Tella, J.L., Scheuerlein, A. and Ricklefs, R.E. 2002. Is cell-mediated immunity related to the evolution of life-history strategies in birds? *Proceedings of the Royal Society of London* B 269: 1059–1066.

ten Kate, K. and Laird, S.A. 1999. *The Commercial Use of Biodiversity: Access to Genetic Resources and Benefit-Sharing.* Earthscan Publications, London.

Théron, A. and Combes, C. 1995. Asynchrony of infection timing, habitat preference, and sympatric speciation of schistosome parasites. *Evolution* 49: 372–375.

Théron, A., Pointier, J.-P., Morand, S., Imbert-Establet, D. and Borel, G. 1992. Long-term dynamics of natural populations of *Schistosoma mansoni* among *Rattus rattus* in patchy environments. *Parasitology* 104: 291–298.

Thompson, J.N. 1987. Symbiont-induced speciation. *Biological Journal of the Linnean Society* 32: 385–393.

Thompson, R.C.A. and Lymbery, A.J. 1996. Genetic variability in parasites and host-parasite interactions. *Parasitology* 112: S7–S22.

Thursz, M.R., Thomas, H.C., Greenwood, B.M. and Hills, A.V.S. 1997. Heterozygote advantage for HLA class-II type in hepatitis B virus infection. *Nature Genetics* 17: 11–12.

Torchin, M.E., Lafferty, K.D. and Kuris, A.M. 2002. Parasites and marine invasions. *Parasitology* 124: S137–S151.

Torchin, M.E., Lafferty, K.D., Dobson, A.P., McKenzie, V.J. and Kuris, A.M. 2003. Introduced species and their missing parasites. *Nature* 421: 628–630.

Trouvé, S., Sasal, P., Jourdane, J., Renaud, F. and Morand, S. 1998. The evolution of life-history traits in parasitic and free-living platyhelminths: a new perspective. *Oecologia* 115: 370–378.

Trowsdale, J., Groves, V. and Arnason, A. 1989. Limited MHC polymorphism in whales. *Imunogenetics* 29: 19–24.

Turner, J.R.G., Gatehouse, C.M. and Corey, C.A. 1987. Does solar energy control organic diversity? Butterflies, moths and the British climate. *Oikos* 48: 195–205.

Valkiunas, G. and Ashford, R.W. 2002. Natural host range is not a valid taxonomic character. *Trends in Parasitology* 18: 528–529.

Van Valen, L.M. 1973. Body size and numbers of plants and animals. *Evolution* 27: 27–35.

Vaughn, C.C. 1997. Regional patterns of mussel species distributions in North American rivers. *Ecography* 20: 107–115.

Vaughn, C.C. and Taylor, C.M. 2000. Macroecology of a host-parasite relationship. *Ecography* 23: 11–20.

Via, S. 2001. Sympatric speciation in animals: the ugly duckling grows up. *Trends in Ecology and Evolution* 16: 381–390.

Vickery, W.L. and Poulin, R. 1998. Parasite extinction and colonization and the evolution of parasite communities. *International Journal for Parasitology* 28: 727–737.

Vickery, W.L. and Poulin, R. 2002. Can helminth community patterns be amplified when transferred by predation from intermediate to definitive hosts? *Journal of Parasitology* 88: 650–656.

Volkov, I., Banavar, J.R., Hubbell, S.P. and Maritan, A. 2003. Neutral theory and relative species abundance in ecology. *Nature* 424: 1035–1037.

Walter, D.E. and Proctor, H.C. 1999. *Mites: Ecology, Evolution and Behaviour.* CAB International, Wallingford, UK.

Walther, B.A., Cotgreave, P., Price, R.D., Gregory, R.D. and Clayton, D.H. 1995. Sampling effort and parasite species richness. *Parasitology Today* 11: 306–310.

Walther, B.A. and Morand, S. 1998. Comparative performance of species richness estimation methods. *Parasitology* 116: 395–405.

Wang, S.W. and Shiao, M.S. 2000. Pharmacological functions of Chinese medicinal fungus *Cordyceps sinensis* and related species. *Journal of Food and Drug Analysis* 8: 248–257.

Warén, A. 1984. A generic revision of the family Eulimidae (Gastropoda, Prosobranchia). *Journal of Molluscan Studies* 13 (Suppl.): 1–96.

Warwick, R.M. and Clarke, K.R. 2001. Practical measures of marine biodiversity based on relatedness of species. *Oceanography and Marine Biology Annual Review* 39: 207–231.

Waters, A.P., Higgins, D.G. and McCutchan, T.F. 1991. *Plasmodium falciparum* appears to have arisen as a result of lateral transfer between avian and human hosts. *Proceedings of the National Academy of Sciences U.S.A.* 88: 3140–3144.

Watters, G.T. 1992. Unionids, fishes, and the species-area curve. *Journal of Biogeography* 19: 481–490.

Watve, M.G. and Sukumar, R. 1995. Parasite abundance and diversity in mammals: correlates with host ecology. *Proceedings of the National Academy of Sciences U.S.A.* 92: 8945–8949.

Webster, J.M., Chen, G. and Li, J. 1998. Parasitic worms: an ally in the war against the superbugs. *Parasitology Today* 14: 161–163.

Wedekind, C., Chapuisat, M., Macas, E. and Rülicke, T. 1996. Non-random fertilization in mice correlates with the MHC and something else. *Heredity* 77: 40–409.

Wedekind, C. and Folstad, I. 1994. Adaptive or non-adaptive immunosuppression by sex hormones? *American Naturalist* 143: 936–938.

Wedekind, C. and Füri, S. 1997. Body odour preferences in men and women: do they aim for specific MHC combination or simply heterozygosity? *Proceedings of the Royal Society of London* B 264: 1471–1479.

Weeks, A.R., Reynolds, K.T. and Hoffmann, A.A. 2002. *Wolbachia* dynamics and host effects: what has (and has not) been demonstrated? *Trends in Ecology and Evolution* 17: 257–262.

Wegner, K.M., Reusch, T.B.H. and Kalbe, M. 2003. Multiple parasites are driving major histocompatibility complex polymorphism in the wild. *Journal of Evolutionary Biology* 16: 224–232.

Whittington, I.D. 1998. Diversity 'down under': monogeneans in the Antipodes (Australia) with a prediction of monogenean biodiversity worldwide. *International Journal for Parasitology* 28: 1481–1493.

Wilckens, T. and de Rijk, R. 1997. Glucorticoids and immune function: unknown dimensions and new frontiers. *Immunology Today* 18: 418–424.

Willig, M.R., Kaufman, D.M. and Stevens, R.D. 2003. Latitudinal gradients of biodiversity: pattern, process, scale and synthesis. *Annual Review of Ecology, Evolution, and Systematics* 34: 273–309.

Wilson, E.O. 1992. *The Diversity of Life*. Harvard University Press, Harvard.

Wilson, E.O. 2003. The encyclopedia of life. *Trends in Ecology and Evolution* 18: 77–80.

Wilson, K., Bjørnstad, O.N., Dobson, A.P., Merler, S., Poglayen, G., Randolph, S.E., Read, A.F. and Skorping, A. 2002. Heterogeneities in macroparasite infections: patterns and processes. In: *The Ecology of Wildlife Diseases* (Hudson, P.J., Rizzoli, A., Grenfell, B.T., Heesterbeek, H. and Dobson, A.P., editors). Oxford University Press, Oxford, pp. 6–44.

Wilson, N. and Durden, L.A. 2003. Ectoparasites of terrestrial vertebrates inhabiting the Georgia Barrier Islands, USA: an inventory and preliminary biogeographical analysis. *Journal of Biogeography* 30: 1207–1220.

Windsor, D.A. 1998. Most of the species on Earth are parasites. *International Journal for Parasitology* 28: 1939–1941.

Worthen, W.B. 1996. Community composition and nested-subset analyses: basic descriptors for community ecology. *Oikos* 76: 417–426.

Worthen, W.B. and Rohde, K. 1996. Nested subset analyses of colonization-dominated communities: metazoan ectoparasites of marine fishes. *Oikos* 75: 471–478.

Wright, D.H., Currie, D.J. and Maurer, B.A. 1993. Energy supply and patterns of species richness on local and regional scales. In: *Species Diversity in Ecological Communities: Historical and Geographical Perspectives* (Ricklefs, R.E. and Schluter, D., editors). University of Chicago Press, Chicago, pp. 66–74.

Wright, D.H., Patterson, B.D., Mikkelson, G.M., Cutler, A. and Atmar, W. 1998. A comparative analysis of nested subset patterns of species composition. *Oecologia* 113: 1–20.

Yamazaki, K., Beauchamp, G.K., Wysocki, C.J., Bard, J., Thomas, L. and Boyse, E.A. 1994. Discrimination of odor types determined by the major histocompatibility complex among outbred mice. *Proceedings of the National Academy of Sciences U.S.A.* 91: 3735–3738.

Yamazaki, K., Beauchamp, G.K., Wysocki, C.J., Bard, J., Thomas, L. and Boyse, E.A. 1983. Recognition of H-2 types in relation to the blocking of pregnancy in mice. *Science* 221: 186–188.

Yamazaki, K., Boyse, E.A, Thaler, H.T., Mathieson, B.J., Abbott, J., Boyse, J. and Zayas, Z.A. 1976. Control of mating preferences in mice by genes in the major histocompatibility complex. *Journal of Experimental Medicine* 144: 1324–1335.

Zelmer, D.A. and Esch, G.W. 1999. Robust estimation of parasite component community richness. *Journal of Parasitology* 85: 592–594.

Ziegler, A., Ehlers, A., Forbes, S., Trowsdale, J., Uchanska-Ziegler, B., Volz, A., Younger, R. and Beck, S. 2000. Polymorphic olfactory receptor genes and HLA constitute extended haplotypes. In: *Major Histocompatibility Complex: Evolution, Structure, Function* (Kasahara, M., editor). Springer-Verlag, Tokyo, pp. 110–130.

Zietara, M.S. and Lumme, J. 2002. Speciation by host switch and adaptive radiation in a fish parasite genus, *Gyrodactylus* (Monogenea, Gyrodactylidae). *Evolution* 56: 2445–2458.

Zrzavy, J. 2001. The interrelationships of metazoan parasites: a review of phylum- and higher-level hypotheses from recent morphological and molecular phylogenetic analyses. *Folia Parasitologica* 48: 81–103.

Zrzavy, J., Mihulka, S., Kepka, P., Bezdk, A. and Tiet, D. 1998. Phylogeny of the Metazoa based on morphological and 18S ribosomal DNA evidence. *Cladistics* 14: 249–285.

Zuk, M. 1992. The role of parasites in sexual selection: current evidence and future directions. *Advances in the Study of Behavior* 21: 39–68.

Zuk, M. 1996. Disease, endocrine-immune interactions, and sexual selection. *Ecology* 77: 1037–1042.

Zuk, M. and Stoehr, A.M. 2002. Immune defense and host life history. *American Naturalist* 160: S9–S22.

Index

DATE DUE